New Concepts in Polymer Science

Biodegradation and Durability of Materials under the Effect of Microorganisms

New Concepts in Polymer Science

Previous titles in this book series:

Fire Resistant and Thermally Stable Materials Derived from Chlorinated Polyethylene
A.A. Donskoi, M.A. Shashkina and G.E. Zaikov

Polymers and Polymeric Materials for Fiber and Gradient Optics
N. Lekishvili, L. Nadareishvili, G. Zaikov and L. Khananashvili

Polymers Derived from Isobutylene. Synthesis, Properties, Application
Yu.A. Sangalov, K.S. Minsker and G.E. Zaikov

Ecological Aspects of Polymer Flame Retardancy
S.M. Lomakin and G.E. Zaikov

Molecular Dynamics of Additives in Polymers
A.L. Kovarski

Structure and Properties of Conducting Polymer Composites
V.E. Gul'

Interaction of Polymers with Bioactive and Corrosive Media
A.L. Iordanskii, T.E. Rudakova and G.E. Zaikov

Immobilization on Polymers
M.I. Shtilman

Radiation Chemistry of Polymers
V.S. Ivanov

Polymeric Composites
R.B. Seymour

Reactive Oligomers
S.G. Entelis, V.V. Evreinov and A.I. Kuzaev

Diffusion of Electrolytes in Polymers
G.E. Zaikov, A.L. Iordanskii and V.S. Markin

Chemical Physics of Polymer Degradation and Stabilization
N.M Emanuel and A.L. Buchachenko

Of related interest:

Journal of Adhesion Science and Technology
Editor: K.L. Mittal

Designed Monomers and Polymers
Editor-in-Chief: M.K. Mishra

Composite Interfaces
Editor-in-Chief: H. Ishida

New Concepts in Polymer Science

Biodegradation and Durability of Materials under the Effect of Microorganisms

S.A. Semenov*, K.Z. Gumargalieva*
and G.E Zaikov**

* N.N. Semenov Institute of Chemical Physics, Russian Academy of Sciences, Moscow, Russia
** N.M. Emanuel Institute of Biochemical Physics, Russian Academy of Sciences, Russia

CRC Press is an imprint of the
Taylor & Francis Group, an **informa** business

First published 2003 by VSP Publishing

Published 2018 by CRC Press
Taylor & Francis Group
6000 Broken Sound Parkway NW, Suite 300
Boca Raton, FL 33487-2742

First issued in paperback 2019

ISBN 13: 978-0-367-44665-9 (pbk)
ISBN 13: 978-90-6764-388-7 (hbk)

Visit the Taylor & Francis Web site at
http://www.taylorandfrancis.com

and the CRC Press Web site at
http://www.crcpress.com

"God is in details"

Volter, France

"Incomplete is worse than undone"

Alexander V. Suvorov, Generalissimo
of Russia, 18th century

CONTENTS

PREFACE

In recent decades, drastic progress in development of machinery and its wide spreading all over the world have put forward many serious problems before people operating machines and mechanisms. One of them, arisen in the middle of the previous century, was that machines and devices produced in developed countries became disabled rapidly, when appeared under unusual conditions. These unusual conditions were high humidity, temperature and other climatic conditions typical of subtropical, tropical and equatorial zones of the Globe, where the machinery was mainly exported (Korea, Laos, Vietnam, Angola, etc.).

That was the time, when the scientists have recognized that not only metals, wood and fabrics, but organic materials (polymers, rubbers, oils and fuels) are also subject to aggressive impact in hot and humid climate conditions.

However, it has been found that not combinations of temperature and humidity are the most dangerous parameters, which destroy materials fasted than in the moderate climate. Now it is obvious that one of the main hazards to machinery is active growth of various microorganisms (bacteria, fungi, etc.) on them, which consume carbon-containing substances destroying the organics.

The present monograph is devoted to this very problem. The data shown in the book have been collected during long years under the most aggressive climate conditions and have not been published yet. The main aim of these investigations was to find the way of stabilization and increase of durability (i.e. reduction of aging) of organic, polymeric and other materials impacted by microorganisms.

The book is intended for serious investigators as a reference material, for students and postgraduates as a textbook, and for engineers working in the branches, in which materials are subject to active microbiological aging and destruction.

The authors of the monograph would be thankful for any constructive questions and note concerning the problems indicated in the book and aspects associated with biodegradation of materials.

Sergei A. Semenov, Klara Z. Gumargalieva – N.N. Semenov Institute of Chemical Physics, Russian Academy of Sciences
Gennady E. Zaikov – N.M. Emanuel Institute of Biochemical Physics, Russian Academy of Sciences

Introduction

The reliability of technical articles is basically defined by their resistance to the environmental impacts, among which microorganisms represent the natural component (microscopic fungi, bacteria, yeast, etc.). Affecting the technical objects, microorganisms-destructors (biological factor, biological destructors) cause their damages (biological damage, microbiological damage): a change in structural and functional characteristics up to destruction.

Biological destructors (biodestructors) are capable of rapid adaptation to various environmental conditions, materials (as the nutritious sources), and protection measures. Moreover, the initial resistance of materials to a biological factor (biofactor), built-in during production, can be abruptly reduced during operation. Therewith, almost all known materials are subject to biodamaging, the injury from which is estimated as 2 – 3% vol. of total industrial production.

At the same time, the features and regularities of the biofactor impact are studied in much less degree than the influence of such non-biological factors as temperature, mechanical stress, light radiation, aggressive media, etc. on materials and articles. At the present time, the great attention of investigators is devoted to environmental and biological components of the problem of machinery biodamaging. The species composition, features and capability of microorganisms to occupy materials are studied, and selection of protection measures is usually carried out empirically. Despite the great amount of activities held in this direction, the means applied do not often provide a sufficient resistance of articles to the effect of microorganisms. In conditions of operation, failures and breakdowns of separate aggregates and systems of aircrafts, ships, motor vehicles and other technical articles and machinery are observed, caused by the microbiological damage of materials.

Low protection efficiency is associated with insufficient level of investigation in the field of the material science aspects of damaging action of microorganisms. There are no quantitative data on the processes of machinery biodamaging under actual conditions of operation and scientifically proved ideas on the mechanism of such damages. There is illustrative information only, which is not often based on the experimental results. No reliable methods of diagnostics and forecasting are elaborated till present.

The problem can be successfully solved by investigating the nature and kinetic regularities of the interaction between materials and biodestructors. These investigations will allow stipulation of scientific and methodological approaches to objective and reliable estimation and forecasting of the microbiological resistance of technical articles and machinery. They will also promote development of materials and structures resistant to biological damages (bioresistant materials), and efficient protection means and methods.

The main goal of the monograph is development of theoretical notions about the mechanism of material damaging by microorganisms under operation conditions and compilation of recommendations to protection of technical articles and machinery. The basic tasks are the following:

- Finding of kinetic regularities, analytical models and quantitative indices for stages of material biodamaging;
- Study of influence of the material properties, features of a microorganism, temperature, humidity, and other factors on interaction between the material and biodestructor;
- Compiling guidelines to determination and forecasting of biological resistance (bioresistance) of the materials, development of protection means and methods, and estimation of their efficiency.

Chapter 1.

Modern ideas about biodamaging of materials and technical articles. Review of references

Biodamaging (biological damaging) represents any change (distortion) of structural and functional characteristics of an object induced by a biological factor [1, 2]. By biological factor organisms or their associations are meant, the influence of which on a technical facility disturbs its operational order. Microorganisms (microscopic fungi, bacteria, and yeast [2 – 4]) are most aggressive to materials and articles. Being the components of the environment and due to specificity of their vital activity, biodestructors are capable of rapid adaptation to various materials and continuously alternating conditions. Practically all materials used in technical articles are subject to the damaging effect of microorganisms which is the microbiological damage [2, 4].

The presence and intensity of the microbiological damage characterizes resistance of technical facilities to microbiological factor (microbiological resistance), which is the property of facilities (material, component or article) to preserve values of indices within the range, given in the regulatory technical documents, during the given time under the effect of a microbiological factor or after it [1].

The significance of the problem of biodamages is outlined by many specialists. For example, the worldwide injury from biodamages in 1950's was estimated as 2% of the total industrial production, and in 1970's it exceeded 5% that gives tens of billion dollars. About a half of the total volume of injury is caused by microbiological damages. In the USA, the injury from corrosion of iron pipes, caused by sulfate-restoring bacteria, is estimated as 2 billion dollars per year [3]. It is assumed that over 50% of all corrosion processes is associated with the action of microorganisms [4].

1.1. MATERIALS SUBJECT TO DAMAGE AND MICROORGANISMS-DESTRUCTORS

Microbiological damages of combustible materials and lubricants (CML) are the most completely studied ones [5 – 11]. It has been found that possessing the ability to enzymatic oxidation of liquid hydrocarbons some species of microorganisms consume them as the source of nutrition. Assimilating such hydrocarbons and affecting them with products of their vital activities, the microorganisms-destructors induce degradation and loss of operational properties of the material. Oil fuels are subject to microbiological damage during both storage and transportation, and operation [12 – 16].

The study of problems associated with the development of microorganisms in fuels has been initiated in the USA simultaneously with development of jet-propelled aviation. It has been shown that accumulation of the products of growth and vital activities of microscopic fungi and bacteria in the fuel systems may cause filter blinding and stoppage of other aggregates, breaks in operation of fuel control equipment sensors, damage of internal protective coatings, and corrosion injuries of materials in fuel supply systems [17 – 22].

It has been found [23, 27] that practically all fuels, including aviation kerosenes, are unstable to the impact of microorganisms and represent a favorable medium for growth of some species of microscopic fungi and bacteria. Growth of the microorganisms may induce accumulation of a great amount of biomass and aggressive products of vital activities (organic acids, amino acids, enzymes, etc.) in the fuel systems. If necessary measures are not undertaken, this will induce breakdowns and disorders in operation of technical articles and machinery.

As reported in ref. [28], intensive microbiological processes proceed in the fuel tanks of sea ships during travels. Microorganisms "process and make bad" from 100 to 900 mg of fuel (diesel oil) per cubic meter daily. Therewith for diesel oil, such index as "the quantity of mechanical admixtures" may by ten and hundred times exceed the permissible level.

As noted in ref. [29], all motor, diesel, petrolatum, spindle, aviation, almost all transmission and insulation oils, and plastic lubricants are unstable to fungi and bacteria. Impacted by these microorganisms, the majority of indices of oil and lubricant properties (viscosity, acid number, resistance to oxidation, etc.) are changed significantly. Corrosion of associates and parts contacting with the damaged oils and lubricants occurs. The cases of breakdowns in operation of oil aggregates due to microbiological damages of active oils are observed [28].

Many authors outline strong dependence of the microbiological resistance of CML of even the same type on the initial raw material and production technology. For example, oils for different purposes produced from Anastasievskaya crude oil are the most stable, and transformer fluid from this oil is considered the one "absolute stable" to microorganisms [29]. Such differences are induced by the features of group and individual hydrocarbon composition of particular material. It has also been found that many sulfur compounds present in a sulfurous crude oil significantly reduce microbiological resistance of lubricants, produced from it. At the same time, nitrogen-containing compounds present in resinous oil fractions are active biocides – the substances which kill microorganisms.

Among quantitative external conditions promoting microbiological damages of CML, the presence of water, mineral admixtures (pollutants) in the material and optimal temperature for development of microorganisms are the characteristic ones.

Bacteria of Pseudomonas, Micrococcus, Mycobacterium, and Desulfovibrio genus, fungi of Cladosporius, Aspergillus, Penicillium, and Chaetomium genus, and yeast of Candidas and Torula genus are typical microorganisms, capable of development in the fuels [2, 4, 29].

Oils and lubricants are most often damaged by the following species of fungi and bacteria: fungi - *Aspergillus niger*, *Penicillium variabile*, *Penicillium chrysogenum*, *Penicillium verrucosum*, and *Scopulariopsis brevicaulis*; bacteria - *Bacillus subtilis*, *Bacillus pumilus*, and *Bacillus licheniformis* [2, 4, 29].

Many works are devoted to the problems of microorganism impact on metals and metal alloys [4, 30, 31]. Resulting vital activities of microorganisms, specific aggressive media and electrochemical concentrated elements are formed on the metal surface, aggressive chemical compounds occur in the surrounding medium (solution) and (or) on the surface, and electrochemical potentials are changed. Fungi and bacteria are capable of initiation and intensification of various types of corrosion.

Many bacteria are able to initiate corrosion of even usually corrosion-proof metals and alloys, for example, copper, lead, etc. [30]. Fungal corrosion is most often observed on technical articles (equipment, devices, structured associates and aggregates), in which metal parts contact with materials promoting development of fungi [32 – 38].

Refs. [39, 40] display results of experimental studies on destructions of aluminum and magnesium alloys induced by *Aspergillus niger* and other fungi. A variety of corrosion damages (separation into layers, pitting, intercrystallite corrosion) is denoted. The fatigue limit of tested samples is reduced to the greater extent than after usual chemical influence.

As indicated in ref. [41], fungi of Aureobasidium, Alternaria, and Stemphylium genus actively corrode aluminum and steel alloys.

Tests on steel, aluminum and copper wires impacted by microscopic fungi [42, 43] have indicated lower resistance of steel samples. *Aspergillus niger*, *Aspergillus amstelodami*, and *Penicillium cyclopium* are the most aggressive cultures. At the same time, *Chactomium globosum* and *Stachybetris atra* induce no noticeable changes. Destruction of copper wires was observed in the case of *Penicillium cyclopium* impact only.

Varnish coatings (VC), polymeric materials, and rubber-technical articles are subject to the negative effect of microorganisms [44, 45]. Their consumption as the source of nutrition and damaging by microorganisms is generally associated with the effect of substances, produced by microscopic fungi and bacteria during their vital activity. Properties of the materials are changed as a result of various chemical reactions: oxidation, restoration, decarboxylation, etherification, hydrolysis, etc. [3].

Chemical composition of a film-forming polymer and physical properties of a cover formed by its film (ability to swelling, hardness, porosity, water repellency, etc.) are of the significant meaning for microbiological resistance of VCs [2]. Depending on the presence of biocide properties, substances used as dyes are able to affect seriously the resistance of VCs to microorganisms. As shown in refs. [45 – 47], iron oxide in the VC composition stimulates growth of microorganisms; titanium dioxide is inert; and zinc oxide decelerates the growth. Among the fillers, asbestos and talcum intensify growth, and calcium carbonate decreases it. Low resistance of some VCs to fungi is associated with the presence of magnesium oxide in them as the filler. This compound is hygroscopic and absorbs water, swells and thus promotes intensive development of microorganisms [48].

Many investigators have outlined high impact of external factors (solar radiation, temperature oscillations and air humidity, surface pollution by dust and salts, action of various gases, etc.) on microbiological resistance of the VCs. These factors promote VC aging and prepare alimentation medium for microorganisms [49 – 56]. Microbiological damages of the VCs are also favored by disorders in cover application processes and service requirements during their use [57, 58]. The study of epoxy enamels (*EP-525*, *EP-567*) applied on the base coat *AK-070* has indicated that the integrated air humidity, temperature and metabolites of fungi are the main factors reducing physicomechanical and protective properties of the VCs. The greatest reduction of the strength characteristics is observed during initial 15 - 30 days during active growth of the fungi [59].

Mould fungi are the main agents of microbiological damaging of the VCs. Bacterial damages are less frequent. Fungi damaging the VCs are mostly of Aspergillus, Penicillium, Fusarium, Trichoderma, Alternaria, Cephalosporium, and Pullularia genus, and bacteria are of Pseudomonas and Flavobacterium genus [2, 53].

Microbiological damages of polymers used in structures of machines and equipment are quite frequent. It has been counted that among the totality of damages, induced by microorganisms, parts from plastics are damaged in 25% of cases [60]. Over 60% of polymeric materials used in technique possess insufficient microbiological resistance [4]. Most often, their microbiological damages occur due to the effect of microscopic fungi, which change color and structure of polymer, and hermetic sealing and strength for thin films.

The existing contradictory information about fungus resistance of various polymers should be noted. Polyethylenes, Polyvinylchloride, fluoroplasts, plastic foams, and poly(vinyl acetate) are considered to be quite resistant materials [61, 62]. At the same time, as reported in refs. [63, 64], nylon and neoprene, polyethylene and Polyvinylchloride are damaged by fungi. As indicated in refs. [65, 66], polyethylene is overgrown by *Neurospora sitophila* fungus. In turn, in ref. [67], this material is assumed as the most stable one, because spores of fungi applied to it have not developed during a long time.

Among microorganisms-degraders of polymers, the following species of fungi are the most typical ones: *Aspergillus wamori*, *Aspergillus niger*, *Aspergillus oryzae*, *Trichoderma sp.*, *Aspergillus amstelodami*, *Aspergillus flavus*, *Shaetomim globosum*, *Trichoderma lignorum*, *Cephalosporum aeremonium*, *Penicillium sp.*, *Rhizopus nigricans*, and *Fusarium roseum* [68 – 71].

Microbiological resistance of rubber-technical articles (RTA) depends on their component composition. Reported in refs. [68 – 75] is low resistance of the main component (rubber) to the impact of microorganisms. Growth of the microorganisms is also promoted by other components (stearine, dibutylphthalate) [76]. Resistance to fungi is also associated with the RTA production process; it depends on the type of vulcanization, in particular [77].

Many investigators associate intensity of microorganisms' growth on RTA with the processes of their aging under the impact of external factors (light, temperature, pressure, ozone, water, etc.) [78, 79]. These factors induce break of macromolecular chains, variations in the composition of separate units, and degradation of the surface layer of the rubber. All these processes form favorable conditions for development of the microorganisms.

Investigations of the microbiological damaging of optical glass show that even insignificant damages of the glass surface cause serious reduction and even complete loss of workability of optical devices.

As indicated in refs. [80, 81], light transmission coefficient is reduced, on average, by 28%, and light dispersion coefficient is increased, on average, by

5.2 times already at weak development of mould fungi on the surface of optical glass (a monocular). Microbiological damages of oculars and prisms of binocular magnifiers and capacitors of microscopes have been detected [82].

The intensity of microorganisms' growth depends on the type (composition) of glass. For example, mould grows faster on crown glasses containing a good spot of silicon dioxide than on flints containing shorter quantity of silicic acid, and the main part of which is composed of heavy metals [82, 83].

Stimulating factors as pollution during production and operation of optical devices, contacts with non-bioresistant parts (washers, lubrication, etc.), from damaged areas of which biodestructors are transported to the glass surface, are sufficient for the occurrence of microbiological damages of the glass [84 – 87].

A miscellaneous species composition of microscopic fungi has been identified on optical parts in various climatic regions: *Aspergillus niger*, *Aspergillus versicolor*, *Aspergillus oryzae*, *Aspergillus flavus*, *Penicillium luteum*, *Penicillium spinulasum*, *Penicillium commune*, *Penicillium citrinum*, *Penicillium frequentens*, *Mucor sp*., etc., totally 23 species [2 – 4].

The works devoted to microbiological damages of electrical and radio articles (ERA) testify about high sensitivity of ERA to vital activity processes of microorganisms. The components with heterogeneous composition of the material and production technology possess different microbiological resistance. Microorganisms developing on any part are capable of inducing damages of articles resistant to their influence, reduce operation parameters and even cause breakdowns of the article operation. As concluded by specialists working in the field of reliability of radio and electronic equipment, 0.5% of equipment breakdowns are induced by impact of the microbiological medium [88].

It is numerously noted about the mould growth on the surface of electrical insulation causes abrupt reduction of its electrical resistance [89 – 91]. The authors note about danger of the mould fungi occurrence on the materials of electronic instruments [92 – 94]. It is also reported that the mould occurred due to inadequate ventilation systems and high humidity causes frequent breakdowns of ERA on atomic missile submarines and above-water crafts [95, 96].

The greatest quantity of microbiological damages in materials and parts of ERA is observed under tropical climate conditions. Anyway, they are observed in all climatic zones [97 – 99].

Investigations carried out at facilities of river and marine fleet, operated (stored) in various climatic zones, have indicated that the following materials are the main focuses of the mould fungi spreading over the radio and electronic equipment: cotton threads, leatherette, rubber, ceramics, fiber, plastics, felting, felt, varnished fabric, etc. Caverns, swellings and cracks occur on plastic boards of the devices, caused by microorganisms. The cases of degradation of insulation materials (varnished fabric, VC, PVC tubes), insulation breakdowns of electric wires and boards, failures of transformers, etc. were observed. Many parts of electric equipment from steel, aluminum and magnesium alloys (St. 10, 45, 30KhGSA, D16T, AMG, AMT, MA-16, etc.) as well as parts with zinc, cadmium, phosphate, and tin coatings are subject to microbiological corrosion.

1.2. INTERACTION BETWEEN MATERIALS AND MICROORGANISMS-DESTRUCTORS

Despite to variety of the processes and reasons inducing occurrence and development of the microbiological damage, they can be reduced to a series of interactions between materials and microorganisms, general for all biodamaging situations. First, the presence of the biodestructor source is required; then they should be transferred from this source to potentially damaged surface (material, part or article). At last, fixing, growth of the microorganism on this surface (or in the volume), and changing of properties of the material contacting with it happen.

The above-mentioned stages of the general process are studied at different extents. Usually, works on problems of biodamages discuss only growth of microorganisms and its effect on the material properties.

Some aspects of microbial cells transfer in the atmosphere, their fixation and growth on various substrates are discussed in the literature on atmospheric, soil and medical microbiology, adhesion of finely dispersed particles, and propagation of biological systems. Study of the effect of aggressive media, to which microorganisms on materials may also be related, is the subject of multiple investigations of the material science.

In this connection, it seems desirable to analyze literary information about the nature, regularities and methodology of studies of the interactions between materials and microorganisms using notions of these interactions, typical of the above-mentioned disciplines.

It is common knowledge that microorganisms are spread all everywhere. They occupy soils, water, air, animal and human organisms [100]. Anyway, the main place of their dwelling is soil, and the species composition and amount of them depend on the soil natural and chemical features. The soil layer from 5 to 15 cm deep is the most saturated with microorganisms. One gram of soils contains up to 10^8 microorganisms [100, 101]. Distribution of various groups of them is defined by the soil type [102]. Penicillium and Aspergillus fungus genus and Bacillus mycoides and Bacillus megaterium bacteria are the microorganisms-cosmopolites for all types of soils.

The composition of microorganisms in the atmospheric air depends on their concentration in soils and water, as well as on the season and meteorological conditions.

Microorganisms are mostly reproduced and spread by spores [103]. Despite the great variety in size and shape, a splendid amount of spores represents unicellular formations of oval or spherical shape with the diameter from 1 to 50 μm.

For the majority of microscopic fungi, spreading of spores in nature is divided into two stages [103]: letting spores free from direct contact with the mother tissue and their wind, insect, etc. spreading. Spores of the ground-living fungi are usually cast away actively, when releasing from the mother tissue. Active release may proceed in different ways by spore ejection: bursting turiscent cells, ejection by water ray, extrusion of swollen mucous mass containing spores through cell walls, etc.

There are fungi, which spores are separated passively. Among passive release of spores, the following types are notable: separation due to gravity force; release due to convective flow effect; spore blow away (deflation); separation by collision with fog or rain drops.

Entering the atmosphere, the spores are spread, their concentration in the air decreases as removed from the source. It is assumed that the basic factors determining spore spreading is the wind speed and the speed of spore falling down in calm air [104, 105]. The speed of falling varies from 0.05 cm/s for small spores (∅ 4.2 μm) to 2.8 cm/s for large elongated spores (29 × 68 μm). For spores of the majority of microscopic fungi, it is 1 cm/s, approximately.

As concluded in refs. [100, 103] devoted to distribution of microorganisms in the air, the application of mathematical statistics methods to quantitative determination of the spore presence in different places in different periods of time is rightful. Analytical equations have been suggested, which allow determination of the spore concentration in a cloud formed by different sources. By type of spore ejection, the sources are divided into three types: instantaneous point source, continuous point source, and continuous linear source.

The amount of spores present in the air is determined by different methods. Sedimentation methods, based on precipitation of the particles due to gravity, are widespread. Sedimentation from the calm air is studied with the help of small boxes with tip-up walls. Sedimentation of spores from natural air flow is studied with the help of glass plates exposed under a tent. Other methods, including forced air flow, are also used. Inertial methods (using inertial filters, various nozzles, weathercocks, pumps, ventilators, vertical cylinders, centrifugal intakes, balloons, etc.) are used for studying the character of distribution and sedimentation of spores in different air flows.

The next stage and the necessary condition of the interaction between microorganisms and materials is transfer of microbial cells to the surface potentially subject to damage. In investigations of the atmospheric microbiology, this stage is usually called the precipitation of spores. By this term the processes are meant, which induce transfer of microorganisms present in the air from suspension to the surface of solids (up to the beginning of fixing to the material surface) or liquids [100, 103, 106].

The following main types of precipitation under natural conditions are known: inertial precipitation, sedimentation, exchange in the interface, turbulent precipitation, precipitation by rain, and electrostatic precipitation. All these types of precipitation or sedimentation are characterized by the ratio between the quantity of spores present in the air layer 1 cm thick above the surface and their quantity, precipitated to the current surface.

Any standard surface (usually the glass one), covered by a medium nutritive for microorganisms, is used in investigations. The quantity of spores entrapped by the surface is calculated in accordance with the common microbiological methods [106 – 108].

The interaction between microorganisms and the material becomes apparent in tight closeness to the surface and a distance of hundreds angstroms and causes cell fixation to the material [109, 110]. In the literature, this stage of biodamaging is denoted by different terms: sorption, adsorption, adhesion, fixing, absorption, etc. At the same time, from positions of modern physics and colloid chemistry, absorption of microbial cells (spores) by solid materials may be considered as the process of adhesion [111, 112]. According to the notions mentioned, two surfaces can be assumed the adhered ones, when any work is required to separate them to the primary state.

The existing data allow consideration of adhesion of microorganisms as the necessary condition and the first stage of the material damaging. As shown in ref. [109], glass destruction by bacteria proceeds at the places of their fixation to the glass surface. It has been found that a contact right-of-way from cellulosomes is formed in *Clostudium thermocellum* thermophilic anaerobe at its contact with cellulose. Cellulose is split in this very right-of-way. It has also been shown [113 – 115] that the cells of *Mycobacterium globiforme* culture are able to adsorb on crystals of steroids (methyl testosterone, cortisone and their acetates). Crystals are dissolved most rapidly in the places of cell fixing.

Microbial adhesion processes are subjects of medical microbiology, virology, ecology, and the soil science (microbiology of soils). Investigations are mainly devoted to interaction of bacteria with various substrates in aqueous media. Problems of microbial adhesion are discussed in detail in monograph [109] and a series of reviews [116 – 125].

Physicochemical ideas and methodology of studying adhesion of finely dispersed particles (dust, powders) of the non-biological origin are developed quite well [111, 112].

Adhesive interaction is determined by the properties of contacting objects and the environment.

The surface of microbial cells is of a complicated morphology (projections, threads, and excrescences). High structure patchiness of its external layer is observed. This layer contains hydrophilic and hydrophobic areas, various functional groups, and areas with positive and negative electric charges [126 – 128]. The cell is capable of alteration during the contact adapting to the substrate and forming adhesive substances, including extracellular ones. Adhesion functions have been detected in heteropolysaccharides, mucosaccharides, and some lipids produced by cells [129 – 131]. Studies of the external layer of plasmatic membrane of *Dictyostelium discoideum* ameba cells have shown that lectins play an important role in aggregation of organisms.

Such properties of the cell provide for the possibility of microorganisms' adhesion to materials possessing various physical and chemical properties (roughness, water repellency, etc.). Such factors as temperature and humidity, pH of aqueous medium, and the presence of various admixtures in it seriously affect this process [119, 120].

Adhesion of microorganisms, as well as finely dispersed particles of the non-biological origin can be stipulated by molecular, chemical, capillary and electric forces [112, 119]. Adhesion is usually divided into two stages. At the initial stage of contact (the first stage), characterized by comparatively weak cell fixation to the surface, physical forces are determinative. Then (the second stage) their character may be changed by occurrence of a chemical component.

Quantitative descriptions of adhesive interaction in separate microorganism-material systems are known, which set the interconnection between the quantity of cells in the substrate (material) and their concentration in the environment [121 – 125]. The ideas of physical and colloid chemistry are successfully applied to the analysis.

Adhesion is commonly characterized by the quantity of cells, fixed at the specific surface, and the force necessary for detaching them. The experimental methods are usually based on two obligatory stages of activities. They are: direct counting of the microbial cells executed by various methods of microscopy (light, luminescent, and electron) and influence of a force field on the adhesion couple (by sloping sample surfaces, centrifuging, vibration detachment, etc.) [112].

Values of adhesion forces of homogeneous finely dispersed particles to the particular surface are different under the same test conditions [109, 112]. That is why the so-called adhesion number is used as the characteristic of adhesion force of such particles. This characteristic is determined as the ratio between the quantity of particles left on the material surface after the force field action of given intensity and the quantity of particles initially present on the surface. Initially, the method of centrifugal detachment has been applied in studies of adhesion of different types of bacteria to glass [109]. It has allowed obtaining of reproducible values of the adhesion number in a broad range of acting force field values.

Adhesive cells begin developing under the conditions favorable for microorganisms, which is characterized by their growth. According to the notions existing in microbiology [132, 133], the growth of microorganisms represents increase of their mass (size of body and/or the quantity of cells) due to consecutively proceeding biochemical reactions of metabolism.

Various aspects of microbiological growth are studied by experimental microbiology [132, 134], genetics [135], physiology, and biotechnology for obtaining various substances in industry [136]. At the present time, growth of microorganisms is the main characteristic of microbiological resistance of materials and efficiency of the existing protection measures [2, 4].

The possibility and intensity of microbiological development are determined by a combination of substrate (material) properties as the nutritive medium for microorganisms and genetic features of the latter, which promote their ability to use the substrate as the nutrition and the source of energy. External conditions (temperature, humidity, pollution, etc.) are also important for the growth [137 - 138].

Microorganisms growing on the material possess rich and labile enzymatic apparatus, which allows their adaptation to many substrates. The enzymes of 6 classes in accordance with the international classification have been found in mycelial fungi [139]. However, oxidoreductases, hydrolases and lipases play a special role in growth on the majority of materials [140 – 142].

Among microorganisms occupying materials based on cellulose, producers of cellulytic enzymes hydrolyzing cellulose to glucose are basic dominants [143 – 149]. Among degraders of hydrocarbon-containing materials, producers of lipases and oxidoreductases are the most active substances. Species producing organic acids, as well as oxidative enzymes – peroxidase and catalase, dominate on materials containing a mineral component. A great quantity of acid-formers, as well as producers of glucose oxidase and poly(phenol oxidase) have been detected on mineral-based construction materials containing organic components [150].

Refs. [45, 151] display regularities of the material structure and the composition action on the growth of microorganisms. For example for polymers, ability to provide microbiological growth is raised with the quantity of carbon atoms in the molecule. It also depends on the degree of substitution, length and chain flexibility between functional groups. As indicated in ref. [152], polyurethanes with the ether bond are more subject to microbiological damages than polyurethanes with the ester bond.

Several main phases (stages) of microorganism development are distinguished in microbiology [132]. They may be of different duration and intensity due to properties of the microorganism-substrate couple. The first stage is the adaptive one (the lag-phase). Duration of it is determined since the moment of material inoculation till reaching the maximum increase of the growth rate. A complex of enzymes necessary for consumption of the given nutritive substrate is synthesized during this period, and irregular accelerated growth of the biomass proceeds that testifies about completion of formation of the appropriate biochemical metabolic mechanisms in cells.

Then a monotonous increase of the biomass at a constant maximal rate proceeds (the exponential phase). This stage lasts up to exhaustion of one of the components of the nutritive medium limiting development of the microorganism and (or) up to formation of metabolites inhibiting its growth. Exhaustion of the nutritive medium and accumulation of metabolites in it lead to a gradual inhibition of metabolic processes, which is accompanied by reduction of the rate and then full stop of the biomass growth (the stationary phase).

At present, numerous equations describing dependence of the growth rate on various factors are suggested [153 – 157]. Many of them are close in the form to the Mono equation [158], which reflects dependence of the enzymatic reaction rate on the concentration of components in the nutritive substrate. Other equations are similar to the one of noncompetitive inhibition of enzymatic reactions. They describe the growth dependence on the concentration of metabolites inhibiting it [159, 160]. Generally, the mentioned models proceed at the exponential stage of growth.

Mathematical models based on solving differential equations of growth, suitable for describing development of various biological objects, are also known. They are based on application of the specific growth rate [161 – 163].

Morphological features and quantitative indices of the biomass, as well as the presence and quantity of substances contained in the cell and (or) metabolites are used as the growth characteristics [2, 4].

Usually, experimental studies of microbiological growth suppose cultivation of microorganisms on the materials and determination of the controlled growth characteristics. Commonly, tests in the field of biodamages provide for a single determination of a characteristic 21-28 and 5-28 days after cultivation of microscopic fungi and bacteria, respectively.

The growth on a material and due to expense of it is accompanied by a change in its properties. These changes proceed in parallel with microbiological growth and often last after its end. They may be caused by a break in the wholeness or pollution of the material due to overgrowth of the microbial bodies. Primarily, changes in properties of the materials at their contact with a biodestructor are associated, quite frequently without any experimental grounds, with the action of organic acids and enzymes produced by the microorganisms [2, 4]. The fullest study of acid formation by microorganisms-destructors has been carried out in works [144, 164 – 166]. Acetic, propionic, butyric, fumaric, succinic, malic, citric, tartaric, gluconic, and oxalic acids have been detected in fungal metabolites. These acids are capable of changing physicomechanical characteristics of the polymers, varnishes and optical glass and causing corrosion of metals [4].

Refs. [144, 145, 167 – 170] display the experimental confirmation of the ability of some enzymatic complexes to degrade materials, such as polymers of animal origin, oil refinement products, and cellulose-containing materials. Regularities in changing properties of these materials by microorganisms and the features of metal corrosion under the effect of a biofactor are studied in detail [2, 4].

A technique for studying the action of aggressive media on materials, to which organic acids, enzymes and other chemical compounds produced by microorganisms could be related, is highly developed [171 – 177]. It is agreed to subdivide aggressive media into two groups with regard to the type of interaction with materials: chemically and physically active ones. The influence of every of these two groups of media frequently induces the same effect. It may be a variation of strength, dielectric properties or the material structure. However, regularities defining such variations depend on the type of the medium effect on the material.

The effect of physically active media is associated with adsorption of molecules of the medium promoting variations of surface properties of the material or caused by swelling (although insignificant) and consecutive plasticization of the material. The effect of physical active media can be of both reversible and irreversible type.

Chemically active media induce irreversible chemical degradation of nonmetal materials and corrosion of metals.

A series of analytical models for aggressive media interaction with solids has been elaborated [173, 175].

The influence of microorganisms on the material properties are characterized by the value of relative, compared with the initial, variation of one index or another of the properties. Experimental studies of this stage of microbiological damaging are performed once after a definite time of biodestructor cultivation on the material. Controlled index of the properties is determined by the methods of analysis known in the material science [2, 4].

Hence, the data present in the literature testify that the composition, biosynthetic properties of microorganisms-destructors, their ability to propagate in technical materials and cause damages of articles are studied quite well. Simultaneously, means and methods used are often low-efficient. Successful solution of the problem is associated in many respects with development of notions about the nature of actual microbiological damaging of the materials and the presence of objective quantitative information about regularities of its occurrence and proceeding.

However, at present, the mechanism of biodamaging is usually considered from positions of biochemical transformations of the material, caused by biodestructor and providing a possibility of its (material) assimilation by this biodestructor as the source of nutritive substances. Quantitative data on this process are absent. Initial interactions of the material with microorganisms, present in the environment, which forego the damage itself, are not studied. The question about the reasons of variations in the material properties induced by biodestructors and the role of compounds produced by microorganisms (metabolites) and other (non-biological) factors of the environment in these variations is not clear yet.

This analysis of the modern state of proceedings on the problem allows a conclusion about availability of applying formal-kinetic ideas of the mechanism of microbiological damaging of materials to the actual process. Such ideas suppose consideration of the studied process as the aggregate of several stages, proceeding of which separately obey a kinetic law and an analytical model reflecting their mechanisms.

Data present in the literature allow a supposition that biodamaging can be presented by three stages of material interaction with a microorganism: 1 – fixation (adhesion); 2 – biodestructor growth on the material; 3 – variations of properties of the material. Therewith, the nature and quantitative regularities of each of these stages should be studied with the help of well-developed ideas and methodological approaches used in the study of appropriate interactions by such scientific disciplines as atmospheric, soil and medical microbiology, adhesion of finely dispersed particles, and chemical resistance of materials.

Such investigations will give an opportunity to substantiate scientific and methodological approaches to solving the entire complex of practically important problems of increasing microbiological resistance of technical articles.

Chapter 2.

General investigation technique

Adhesion and growth of microorganisms and variations of the material properties were studied on the basis of experimental determination of characteristics which allow ascertaining the fact of proceeding and estimate quantitatively intensity of the processes mentioned.

The characteristics were determined with the help of microbiological, biochemical and physicochemical methods of analysis of properties, composition and structure of materials and microorganisms, as well as the products of their interaction. The following methods were used: light and luminescent microscopy, sowing to nutritive media, centrifugal detachment, radiometry, ultraviolet, infrared, luminescent spectral and photometry, electrochemical analysis, thin-layer (surface) chromatography, mechanical, electrical and other tests of the material properties. Characteristics and methods used are discussed in Chapter 3.

Processes of microbiological damaging of operated technical facilities (materials, parts, aggregates, systems of articles) were revealed with the help of a technique achieved by the authors (Chapter 3). It provides for analysis of sediments (pollutants), detected on these objects, as well as damage products and damaged objects themselves. Hence, a complex of characteristics for interactions between microorganisms and materials is determined.

This method helps in estimation of microbiological ability to damage of the most various technical articles: aircrafts and helicopters, motor vehicles, etc. The results obtained allowed substantiation of the choice of materials and microorganisms for future investigations, interactions between which induce microbiological damage of technical parts and aggregates to the utmost telling upon its workability and technical conditions. These materials and microorganisms are shown in Tables 2.1 and 2.2.

All studied materials are industrially produced. PMMA film 0.1 mm thick was produced by casting from solution of organic glass SO-95 in an organic solvent according to the method, described in ref. [181]. Varnished fabric plates were made by applying varnish ET-959 on a cotton support [182]. The samples were used, the shape, size and the method of preparation (from the initial material) of which were stipulated by the method of determining the characteristics of studied process, controlled in the experiment.

Table 2.1

Materials used in tests

Material, trademark, notation	GOST, TU[1]	Characteristics of material samples
Low density polyethylene, trademark 10812-20, (PE)	GOST 16337-70	Film 0.1 mm thick
Poly(methyl methacrylate), organic glass SO-95, (PMMA)	GOST 10667-74	Film 0.1 mm thick
Polyvinylchloride plasticate, I-40-13, (PVC)	GOST 5960-72	Plate 0.6 mm thick, BPVL wire cover
Varnished fabric, (VF)	Cotton base, varnish ET-959, TU-6-10-691-74	Plate 0.8 mm thick, BPVL wire cover
Cellulose triacetate, (CTA)	GOST 7730-74	Film 0.1 mm thick
Polysulfide sealer U-30MES-5	TU-38-10-5462-80	Film 0.3 mm thick
Epoxy resin, (ER)	GOST 21505-73	Plate 0.3 mm thick
Poly(ethylene terephthalate), (PET)	TU 605-1454-71	Film 0.15 mm thick
Glass fiber laminate STK/EP, (GFL)	TU 16.503.055-75	Plate 3 mm thick
Cellulose film, (CF)	GOST 7730-74	Film 0.05 mm thick
Cotton thread of tarpaulin fabric, marking 11205	GOST 15530-76	Threads 0.5 mm in diameter (density 1.2×10^{-3} g/cm^3)
Oil distillate fuel, TS-1	GOST 10227-86	1st category fuel
Pressure oil, MN-7,5U	TU 38101722-78	1st category oil
Aluminum alloy, D-16	GOST 4784-65	Plate 2 mm thick
Steel, St.3	OST 23.4.122-77	Plate 3 mm thick

Strains of microorganisms preserved in a specialized collection were used. Spore radii (shown in Table 2.2) were accepted equal the mean equivalent sphere radius, equal by the surface square to a geometrical body (cylinder, ellipsoid, sphere, etc.), to which each spore of the current species of the microorganism is the most similar by shape and size. Values of such equivalent radii were calculated based on direct determination (taking a microscope image) of the linear size and shape of 100 – 120 spores and the normal law of distribution of its (equivalent radius) value, established during these preliminary investigations for every species of microorganisms.

---

[1] Russian state standards.

Table 2.2

Microorganisms – material destructors used in tests

Species of microorganism	Damaged material	Radius of spores, μm	Density of spores, g/cm^3
Aspergillus niger v. Tieghem	Polymeric materials, metals, oils and lubricants	4.2 ± 1.8	1.12 ± 0.05
Paecilomyces varioti Baimer	Polymeric materials, oils and lubricants	3.6 ± 1.3	1.08 ± 0.08
Penicillium chrysogenum Thom	Polymeric materials, fuel, metals	2.6 ± 1.5	1.13 ± 0.03
Aspergillus flavus Link	Polymeric materials	2.3 ± 1.6	1.14 ± 0.07
Penicillium cyclopium Westling	Polymeric materials	1.6 ± 0.8	1.17 ± 0.05
Aspergillus terreus var aureus Thom and Raper	Polymeric materials	1.2 ± 0.6	1.12 ± 0.09
Cladosporium resinae Lindau de Vries f. Avellaneum	Fuels	2.8 ± 1.3	1.05 ± 0.04
Pseudomonas aerugenosa Schroeter	Fuels	Non-spore-forming	
Bacillus subtilis Ehrenberg	Oils and lubricants, metals	0.7 ± 0.3	1.01 ± 0.07

Quantitative regularities of the microbiological damaging of technical facilities were investigated in laboratory experiments rendering (modeling) actual processes of adhesion, growth of microbial cells, and variations of the material properties. Such experiments included execution of the following operations: preparing samples of materials and microorganism cultures, inspiration of the samples with microbial cells, and incubation of inspired samples (cultivation of microorganisms on the materials) under definite conditions (temperature, humidity, etc.). The above-mentioned operations were executed with the help of the known modes and methods regulated by the state standards to laboratory tests of resistance of appropriate materials to the action of microorganisms [183 – 186].

After a definite time of incubation, a part of samples were removed from tests and one quantitative characteristic or another of the studied process was determined using an appropriate method of analysis. Characteristics were determined regularly during the tests. Such kinetic approach allowed quantitative description of the process at any particular moment, determination of its rate and features of the process proceeding.

The technique for adhesive interaction study [187] provided for sedimentation of fungal spores in calm air to the surface of material samples, incubation of inspired samples and determination of quantitative characteristics of the process – the adhesion number. This number represents the ratio between the quantity of spores remained on the surface after a force field action to that, initially present on the surface. At the initial stage of experiment, directly during precipitation of spores, the effect of force field of various intensities was provided by tilting the sample surface relative to the direction of the spore motion. Then after some time of contact (~ 1 hour) between adhering objects, the detachment force was created by the centrifugation method.

Experimental application of the simplest transportation mode (sedimentation from calm air) of microbial cells towards the material surface allowed neglecting the influence of such factors as speed and type of air flow, and the features of flowing over the sample on adhesion. Moreover, the sedimentation mechanism of transportation fully reflects the actual interaction of microorganisms with the surfaces of internal volumes of articles, materials and parts, present in storages and, frequently, in the open air (if the lamellar air layer at the earth surface is significant).

Spores were precipitated to samples with the help of special chamber, made from steel, equipped with clips moving up and down for placing samples at the current angle to the direction of falling spares, as well as with two removable cassettes at the lower panel. Petri dishes filled with water, designed for determination of the quantity of precipitating spores, are located in the cassettes. For spore dispersion, rubber membranes are placed in the upper part of the chamber. The membranes are connected with one another and with the air input bulb by thin glass tubes. At pulse air input, the membrane oscillates and disperses spores present on it, which then precipitate on the material samples and Petri dishes filled with water.

Specific expense of spores (the quantity of spores crossing the specific square of horizontal cross-section of the chamber) and the adhesion number at various sample tilts were calculated on the basis of spore quantities precipitated on Petri dishes and the samples. Time for spores to reach the sample surface and time for completing precipitation were measured simultaneously.

After that the samples with precipitated spores were centrifuged in air.

It has been determined in preliminary investigations that if the force field intensity is changed to values providing detachment of not more than 15 – 18% of spores, the adhesion number varies within the range of experimental error. A significant experimental error is also observed in the range of the force field intensity causing detachment of, at least, 85% of cells. Consequently, true determination of adhesion numbers by the centrifugation method is possible in the range of 0.2 – 0.85. The authors have limited their experimental studies by this range.

The technique for microbiological propagation study [184, 187] has provided for cultivation of microorganisms on materials with regular determination of quantitative characteristics of the growth – dry specific biomass representing the quantity of microorganisms' dry mass on the specific square (or in the specific volume) of the material surface (or volume), inspired by the microbial cells. This characteristic was determined by gravimetric, chemoluminescent and radiometric methods (Chapter 3). Main indices for the test conditions are shown in Table 2.3.

Table 2.3

Facilities and conditions of tests on microorganism propagation on materials

Material, microorganism, GOST of the method for testing resistance to microorganism action	Solution for inspiration, method, and concentration of microbial cells on the sample	Conditions of microorganism cultivation on the material, method of specific dry biomass determination
Polyethylene, poly(methyl methacrylate), Polyvinylchloride cellulose triacetate, varnished fabric. *Aspergillus niger*, *Penicillium chrysogenum*, *Penicillium cyclopium*, *Paecilomyces varioti* GOST 9.049-91	Chapek-Dox medium without saccharose. Uniform application of fixed amount of spore suspension in aqueous-mineral solution by a micropipette, 10^6 spores per cm^2 of the sample surface	Surface cultivation in the quiescent state. Radiometric (for PE, PMMA, CTA), chemoluminescent (PVC), gravimetric (VF)
Fuel TS-1. *Cladosporium resinae*, *Pseudomonas aerugenosa*, GOST 9.023-74	Aqueous-mineral solution. Contact with aqueous-mineral solution containing 10^6 spores per cm^2 of the fuel – aqueous-mineral phase boundary	Deep cultivations under immovable conditions (Cladosporium) and at continuous mixing (Pseudomonas). Gravimetric
Oil MN-75U. *Bacillus subtilis*, GOST 9.082-77	Tap water. Contact with tap water containing 10^6 bacterial cells per cm^2 of the oil-water phase boundary	Deep cultivations at continuous mixing. Chemoluminescent

Note: The Chapek-Dox medium represents water solution of the following composition (mass parts): KH_2PO_4 – 0.1%; $MgSO_4$ – 0.05%; $NaNO_3$ – 0.2%; KCl – 0.05%; $FeSO_4$ – 0.001%; saccharose – 3%; aqueous-mineral solution is of the following composition (mass parts): KNO_3 – 0.2%; $MgSO_4$ – 0.04%; KH_2PO_4 – 0.03%; $NaNO_3$ – 0.07%.

Fungi and bacteria were cultivated on materials in the presence of a series of mineral components necessary for microorganism growth, specially injected with microbial cells or contained in tap water (in tests on oil MN-7,5U). Such test mode simulates the presence of mineral pollutions that reflects most properly the actual conditions of machinery operation.

Basing on preliminary investigations, primary sample selection for biomass determination was executed after 10-12 hours of microscopic fungi cultivation on them, and after 3 – 8 hours of bacteria cultivation. As a rule, the time period between samplings did not exceed 3 – 4 hours for bacteria and 1 – 3 days for fungi.

The technique for experimental studies of variations in the properties of materials affected by microorganisms [187] provided for cultivation of microorganisms on samples with regular determination of the property index under control. In some cases, material structure and composition were also determined. The main studied material-microorganism couples, indices of material properties controlled during experiments and methods of determination are shown in Table 2.4.

Table 2.4

Controllable indices of material properties and methods of their determination

Material-microorganism interacting couple	Controlled index of material properties	Method of controlled index determination
Poly(methyl methacrylate), *Aspergillus niger*	Induced elasticity limit, σ. Durability, τ_d	Elongation at constant rate of 0.5 – 5 mm/min. Elongation under constant stress of 3 – 35 MPa
Polyvinylchloride plasticate, *Aspergillus niger*	Electrical insulation resistance, *R*	Application of 1,000 V direct electric voltage
Varnished fabric, *Aspergillus niger*	Electrical insulation resistance, *R*	Application of 1,000 V direct electric voltage
Cotton thread, *Aspergillus niger*	Stress at break, σ	Elongation at constant rate of 5 mm/min
Polysulfide hermetic sealer, *Cladosporium resinae*	Induced elasticity limit, σ	Elongation at constant rate of 0.5 mm/min
Fuels, *Cladosporium resinae*	Quantity of mechanical admixtures	Weighing of dried insoluble admixtures
Oils, *Bacillus subtilis*	Quantity of mechanical admixtures	Weighing of dried insoluble admixtures
Aluminum alloy, *Aspergillus niger*	Corrosion depth, *h*	Microscope study of microsections
Steel alloy, *Bacillus subtilis*	Corrosion depth, *h*	Microscope study of microsections

Indices of the properties were determined according to three main modes: 1 – not removing biomass, which occurred on the samples during cultivation of microorganisms; 2 – after biomass removal; 3 – after biomass removal and sample conditioning under standard (normal) conditions. Such modes most fully reflect the features of the microorganisms' effect on parts and articles during operation of machinery.

Experimental data were analyzed using mathematical statistics methods and mathematical modeling of studied processes.

Chapter 3.

Process characteristics and features of technical materials damaging by microorganisms under operation conditions

---

Analysis of the modern ideas on microbiological damages of technical facilities has formed grounds to a suggestion that scientifically proved solution of practically important tasks for providing microbiological resistance of the articles should be based on consideration of the actual interaction of microorganisms-destructors with the materials dividing the process into three main stages, associated with one another. They are: 1 – adhesion of microbial cells transferred to the material surface; 2 – growth of microbial cells on the material surface; 3 – variations in material properties under the impact of the microorganisms.

Realization of such stage-by-stage methodological approach has stipulated for the necessity of surveying for objective characteristics of every mentioned stage (process) and the methods of their experimental determination.

The present Chapter shows results of such investigations. A technique for revealing microbiological damages of the articles are substantiated. Simultaneously, the main results of its use for estimating ability to microorganism damaging of operated machinery are discussed.

3.1. CHARACTERISTICS AND DETERMINATION METHODS OF THE INTERACTION BETWEEN MICROORGANISMS-DESTRUCTORS AND MATERIALS

By the functional designation, characteristics of adhesion, growth of microorganisms, variations in material properties and experimental methods of their obtaining must provide for a possibility of executing investigations provided in the work: quantitative regularities, determination techniques, means and methods of increasing microbiological resistance of the technical facilities.

In this connection, characteristics allowing both detection of the presence and quantitative estimation of the process intensity of microbiological damaging are required. The methods used for their determination must be highly sensitive, specific and, taking into account practical trend of the current investigations, should be realized with the help of quite simple equipment and accessible reagents.

Selection of the particular characteristics and methods was based on the data from the literature on experimental study of the properties, composition and structure of microbiological objects and technical materials participating in the considered processes, as well as specificity of changes in conditions of the microorganism-material system, stipulated by proceeding of each of these processes.

Many of primarily selected characteristics and methods were developed for different purposes: studies of vegetation, animal tissues, bacteria, microscopic fungi, as well as aging of polymers and corrosion of metals. That is why they required estimation of a possibility to be applied to investigations of microbiological damaging of the technical materials.

For this purpose, sediments (pollutions) observed on the surface (or in the volume) of materials were analyzed, which occurred during laboratory modeling of the processes and detected under actual operation conditions of the technical facilities, as well as materials themselves, on which these sediments were present.

The method and investigation objects (materials and species of microorganisms) mentioned in Chapter 2 were used in modeling processes of microbiological damaging in laboratory. Samples of the materials were placed in a special chamber (Chapter 2), where spores of microscopic fungi and bacteria were sprayed. Subsiding in immovable air, they adhered to the sample surfaces. At this stage of tests, characteristics and methods of determination of the adhesive interaction were approved. Other samples with adhered spores were incubated under conditions, favorable for growth of microorganisms-destructors. After a definite time of incubation, the samples with sediments (masses of growth and vital activity products of microorganisms) formed on the surfaces (or in the volumes) were removed from tests, and various characteristics of microbiological growth and variations of materials' properties were determined with the help of chosen methods of analysis.

Sediments and appropriate materials sampled from the articles directly under operation conditions were also studied. Hereinafter, such objects will be called the actual samples. The feature of the actual sediments is their complex and usually unknown composition. Microorganisms and various atmospheric and technological admixtures, products of metal corrosion, polymer aging, oxidation of combustible materials and lubricants, etc. may occur in them [189,

190]. In this case, pre-history of the material damaging (sediment formation, property variations) is also unknown, including the species composition of microorganisms-destructors participating in the process.

The majority of selected methods were modified during investigations with regard to the features of studied objects and processes. Objectiveness and reliability of every parameter and method were estimated by comparing obtained results with the help of various methods of analysis.

The parameters and methods selected are shown in Table 3.1 and discussed in refs. [188, 191, 192]. They can be subdivided into two groups: 1 – indicator ones allowing detection of the process presence only; 2 – quantitative ones also providing for an opportunity to determine values of one process parameter or another.

Adhesive interaction induces fixation of microorganism cells (spores) on the material surface. Consequently, to separate these two objects (the cell and the material) a force field influence of a definite intensity is required. The presence of cells on the surface and force for their detaching characterize the initial stage of microbiological damaging.

The fact of presence of microorganisms-destructors' cells on the material represent the indicator parameter of adhesion.

Table 3.1

Parameters of microbiological damaging of materials and methods of their determination

Parameter	Method of analysis, information source	The essence and designation of the method
Adhesive interaction between microorganisms and materials		
Presence of cells (spores) of microorganisms on the material	Microscopic study Light method (magnification ×70, ×400 – ×600) [193 – 195]	Fixation of magnified image of external signs of microorganism cells: microscopic fungi – finely dispersed formations (∅ 1 – 1.2 μm) of various shapes (spherical, ellipsoid, cylindrical, club-shaped); bacteria – the finest movable (and immovable) particles (∅ up to 1 μm), shaped as balls, ball chains, fibrous and snake-like structures, flagella;
	Luminescent method (radiation by UV-light with the wavelength of 360 – 365 nm) [195 – 199]	Microscopic fungi and bacteria – green luminescence of microorganism cells dyed by fluorochromium. Indication of adhesion signs
	Sowing to dense nutritious media [201 – 205]	Detection of microbial cell growth present in the sample in the contact with nutritious medium. Indication of viable microbial cells' adhesion, identification of microorganisms-destructors

Table 3.1 (continued)

Adhesion number	Detachment of a "great quantity of particles" (surface tilting and centrifugal detachment methods) [109, 112]	Determination (by microscopic study, microphotographing, sowing to nutritious media) of the microbial cells' quantity on the material before and after the action of detachment force of the current intensity. Quantitative estimation of adhesion strength of microscopic fungi and bacteria spores to solids
Growth of microorganisms on materials		
Presence and quantity of the biomass	Visual method [206 – 208]	Detection of signs of microscopic fungi and bacteria biomass visible to the unaided eye: bloom shaped as a thin film (sediment) of felt network-twisted growth and slime of various colors on solids; jelly-like slimy mass, pollutions, and films at the phase interface in liquids. Indication of microorganisms' growth.
	Microscopic study: Light method (at magnification of ×70, ×400 – ×600)	Recording of magnified image of external signs of the biomass: Twisting of thin threads branching as a network (spawn hyphas with ∅ 5 – 10 μm), spores of fungi, bacterial cells;
	Luminescent method (under radiation by UV-light with wavelengths of 360 – 365 nm)	green luminescence of spawn hyphas and microbial cells dyed with fluorochromium. Indication of microorganisms' propagation.
	Sowing to dense nutritious media	Detection of microorganism colony growth on a nutritious medium. Indication of viable biomass, identification of microorganisms-destructors
	Gravimetric method [209 – 212]	Weighing of dried mass of microorganisms. Determination of specific dry biomass of microorganisms.
	Radiometric method [213, 214]	Measurements of β-irradiation intensity from radioactive isotope (indicator) accumulated in biomass of microorganisms, cultivated on the material in contact with this indicator. Indicators are: tritium water (polymeric materials) and *n*-octadecane tritium-labeled at an end group (fuels). Intensity of β-irradiation is proportional to the biomass. Indication and determination of the biomass quantity on the surface or in the volume of the material.

Table 3.1 (continued)

Presence and quantity of proteins and albuminous compounds	Biochemical method [215 – 221]	Detection (visual or instrumental) of calorimetric effects, induced by involving proteins contained in the sample into chemical reactions, dying the reaction mixture in different colors:
	Reaction with ninhydrin	Blue – due to interaction between ninhydrin and amino groups of proteins and amino acids;
	Reaction with triphenyltetrazolium chloride	Red-brown – due to oxidation-restoration interaction between formaldehyde and 2,3,5-triphenyltetrazolium chloride in alkaline medium with an albuminous catalyst;
	Reaction with Kumasi dye J-250	Blue – due to formation of the dye-protein complex in acidic solution;
	Loury method	Pale blue – due to interaction between proteins and a complex reagent (Folin's reagent) containing molybdenum and tungsten ions, and phosphoric acid. Dying intensity is proportional to the protein concentration and determined due to optical density of the reaction mixture at the wavelength of 750 nm. Indication of microbial growth, determination of protein concentration in the sample (the Loury method). Biomass corresponded to the protein concentration determined can be calculated.
	Infrared sprectrophotometry [222]	Obtaining IR-spectrum of the sample with absorption bands typical of functional groups of albuminous compounds: 1,660 cm^{-1} (C=O group in amides) and 1,550 cm^{-1} (N-H and C-N groups in amides). Microbial growth is indicated.
	Spectral fluorescent method [223]	Obtaining of sample fluorescent spectrum with the fluorescence band from 340 to 360 nm excited by ultraviolet beams with the wavelength of 320 nm typical of aromatic amino acids. Microbial growth is indicated.
	Thin-later chromatography [224 – 227]	Obtaining of sample chromatograms with location (raise front) of pink (red) spots (when treated by ninhydrin). Microbial growth is indicated.
Presence of pyridoxine (B_6 vitamin)	Spectral fluorescent method [228, 229]	Obtaining of fluorescence spectrum of the sample with the minimum in the range of 385 – 420-nm wavelengths typical of pyridoxine, when excited by ultraviolet light with the wavelength of 320 nm. Microbial growth is indicated.

Table 3.1 (continued)

Presence of carboxylic acids	Spectral fluorescent method [230]	Detection (visual) of bright pale blue, blue-green and yellow-green fluorescence of compounds formed due to fusion (heating up) of di- and tricarboxylic acids, contained in the sample, with resorcine and concentrated sulfuric acid. Excitation by UV-radiation with the wavelength up to 360 nm. Growth of microorganisms-destructors is indicated.
	Thin-layer chromatography [231 – 233]	Obtaining of sample chromatograms with location (raise front) of yellow spots (when treated by bromcresol green) typical of carboxylic acids. Growth of microorganisms-destructors is indicated.
Presence and quantity of adenosine triphosphate (disodium adenosine-5 triphosphate, ATP)	Chemoluminescent method [234 – 236]	Measurement of luminescence intensity with wavelength of 560 nm due to involving ATP present in the sample into enzymatic reaction of substrate (luminopherine) decomposition in the presence of a biocatalyst (glowworm luciferase). Luminescence intensity id proportional to ATP concentration. Indication of propagation (of propagating biomass), quantitative determination of ATP concentration in the sample. The quantity of living (growing) biomass appropriate to detected ATP concentration can be calculated.
Presence of enzymes	Biochemical method [237 – 241]	Detection (visual) of effects of biochemical reactions, catalyzed by the appropriate type of enzymes:
Oxidases		Blue-violet dying due to formazans formation during restoration of tetrazolium salts by oxidases;
Catalases		Gas evolution (bubbles) due to oxygen formation during hydrogen peroxide decomposition by catalase;
Phosphatases		Violet or purple dying due to interaction between sodium phthalate, para-nitrophenyl phosphate and sodium trihydric phosphate in the presence of phosphatase. Growth of microorganisms-destructors is indicated.

Table 3.1 (continued)

Changes in properties of materials impacted by microorganisms		
Indices of material properties (physicochemical, mechanical, etc.)	Determination methods of property indices in accordance with the norms and other technical documents	Determination of changes in indices of material properties after and (or) during influence of a microorganism on it. Indication of presence and quantitative determination of influence of microorganisms-destructors on the material properties.

This parameter is reliably determined by the methods of light or luminescent microscopy (Table 3.1) only in laboratories and for microorganisms of the known species composition. The presence of adhered viable microbial cells can be detected by sample incubation in contact with a nutritious medium, which composition considers type of the studied material and species features of possible microorganisms-destructors (the method of sowing to nutritious media).

The adhesion number is the quantitative parameter of the process. Its value represents the ratio between the quantities of cells, remained on the material surface after the influence of particular detaching force and the initial one, present on the surface before the force field effect. This parameter is obtained by the so-called methods of detachment of a large quantity of particles.

In investigations of actual samples, determination of adhesion parameters displays several features. It has been found that the cells (spores) of microorganisms and some non-biological components of sediments often possess similar external signs, indistinguishable by microscopic studies. When luminescent microscopy is used, continuous luminescent background often occurs in the field of vision. Its occurrence is probably associated with the luminescence of non-biological objects present in the sediments. In this case, obtaining of reliable results requires application of special substances – background killers which is not always possible and makes analysis significantly complicated.

Sowing to dense nutritious media usually allows estimation (by morphology of the colonies) of the variety of viable microbial cells present in the actual sample. To ascertain their belonging to microorganisms-destructors, additional investigations are required. In the general case, such investigations provide for experimental estimation of aggressiveness of ascertained species of microorganisms to the material. For this purpose, every species of the cells (preliminarily purified from other species) is cultivated on the material with consequent determination of microbial growth parameters and changes in material properties, shown in Table 3.1 (or quantitative indices as shown in Chapters 4, 5, and 6). Hence, work techniques and methods, analogous to the standard ones applied to laboratory estimation of microbiological resistance of technical facilities, are used. Then if required, microorganisms aggressive to the material (microorganisms-destructors), detected in the above-described manner, are identified (to genus or species) by common microbiological methods [106, 134, 242 – 244].

It is difficult to make quantitative estimation of adhesive process during direct investigation of actual samples. Such estimation can be made by reproducing the process under the laboratory conditions. In this case, material samples and microbial cells (spores) picked from the studied (damaged) article are used in the model test on determination of the adhesion number (Table 3.1). If possible, test conditions must simulate the actual ones.

Growth of microorganisms can be characterized by mass of their microbial body and cells (the biomass), as well as by substances contained in them and excreted during growth.

To indicate growth, biomass is determined visually by external signs and by methods of light and luminescent microscopy, as well as by sowing to selective nutritious media. Indicator parameters of growth are also substances produced by microorganisms: albuminous compounds (determined by biochemical and IR-spectrometric methods), amino and carboxylic acids (spectrofluorescent and chromatographic methods), B_6 vitamin (spectrofluorescent method), adenosine triphosphate (chemoluminescent method), and enzymes (biochemical methods).

Some of the mentioned parameters allow ascertaining not only the fact of microbial growth, but also some additional information about this process and microorganisms participating in it. For example, adenosine triphosphate (ATP) is produced by microorganisms during vital activities only and is absent in cells, which terminated development, as well as in ones in the spore form [234 – 238]. That is why detection of this compound in the sample testifies the presence of growing (developing) biomass in it.

Carboxylic acids and enzymes (oxidases, catalases and phosphatases) are compounds produced by microorganisms which are the most aggressive to materials. Hence, according to the point of view present in refs. [243, 245], the mentioned enzymes allow microscopic fungi to consume technical materials as the source of carbon supplement. The presence of such substances in a sample testifies about microbial growth, as well as characterizes degrading properties of growing microorganisms in relation to materials.

Quantitative parameters of growth (Table 3.1) are biomass determined by gravimetric and radiometric methods, concentration of albuminous compounds (the Loury method), and adenosine triphosphate (chemoluminescent method). It has been found that concentration of each mentioned substance is proportional to biomass of microorganisms present in the sample. If required, this allows execution of proper comparable analysis of growth processes, experimental data on which are obtained using any of these quantitative parameters. Hence, selection of a parameter and method of its determination is caused by the features of growth of the current species of microorganisms on a particular material. If the value of biomass is comparatively high (over 1 mg/cm^2) and its full detachment from the material is technically available, the gravimetric method can be used. Parameters determined by radiometric, biochemical (the Loury method) or chemoluminescent method are used for quantitative estimation of growth conditioning on formation of the biomass microquantities (below 0.1 mg/cm^2).

Analysis of determination results of microbial growth parameters in real samples shows that external signs of the biomass visible to unaided eye are very infrequent. In some cases, application of magnifiers and light microscopy allows proper determination of this parameter (biomass). However, mycelium of fungi and cells (spores), observed in this case, often possess structure and shape similar to the components of non-biological sediments. Similar to studies of real adhesion, application of luminescent microscopy often requires selection of background luminescence killers, which is caused by features of the material and multi-component composition of the actual sample.

Biomass indication by the method of sowing to selective nutritious media allows detection of viable microbial cells. As mentioned above, this method is the integral part of activities carried out for studying actual samples and aimed at ascertaining belonging of microorganisms present in them (samples) to destructors of materials and identifying their genus or species. However, the fact of presence of such microorganisms testifies unambiguously about adhesive interaction only, but does not allow judging about proceeding of the growth. In such cases, other parameters of microbiological growth are not detected on surfaces or in volumes of samples, inspired by microorganisms-destructors.

Results of detecting substances produced by microorganisms in the real samples are often confirmed by data on determination of the growth parameters by independent methods. However, non-microbiological components frequently present in such samples are capable of unpredictable changing calorimetric effects of biochemical reactions, IR-spectra, spectra of fluorescence and chromatograms obtained in appropriated analyses. Moreover, there were cases, when the real samples contained biological objects, which origin was not associated with the microbiological growth (plant pollen, insects, etc.). Investigation of such objects gives results analogous to those obtained in the presence of metabolites of microorganisms.

Such characteristic compounds as adenosine triphosphate, enzymatic complexes and appropriate methods of their determination are less sensitive to non-microbiological components present in the real samples. At the same time, no adenosine triphosphate and enzymes were detected in some samples containing fragments of fungal mycelium, spores, and albuminous compounds. This result gives grounds to a supposition that the microbial growth on the material has finished, and living (growing) cells (spores) are absent or their quantity is below sensitivity of analytical methods used.

Quantitative estimation of the microbiological growth during study of the real samples can be executed using radiometric methods (determination of the biomass), chemoluminescent and biochemical methods (determination of adenosine triphosphate and albuminous compound concentrations, respectively). Hence, application of the radiometric methods is limited by the necessity of providing quite full detachment of microbiological objects from other components present in the real sample, as well as selection of control samples for the purpose of obtaining the background level of radioactivity, accumulated by the material.

Variations of material properties induced by microorganisms are characterized by appropriate indices of properties, measured after or during contact with the products of microbial growth. Standard or specially developed methods of index determination are used. Due to their selection, the process study can be of indicator (for example, if material appearance is the parameter, and the method of its estimation is visual checking) and quantitative type (mechanical, physicomechanical and other indices of properties determined by appropriate instrumental methods).

It has been found experimentally that the following indices of properties are sensitive to the action of microorganisms: stress at break (induced elasticity limit) – for poly(methyl methacrylate), poly(vinyl chloride), polyethylene, Capron, cellulose triacetate, cotton and linen fabrics, hermetic sealer U-30MES-5; electric resistance for varnished fabric, poly(vinyl chloride), fiberglass laminate; plasticizer concentration in poly(vinyl chloride) for insulation; optical transmission for optical glasses; acidity and concentration of mechanical admixtures in fuels and oils; corrosion damage degree for steel and aluminum alloys.

Studies of conditions of operated materials possessing signs of microbiological growth often allowed detection of changes in appearance, strength, electrical resistance, optical properties, component composition and other indices of properties. At the same time, no features of the changes mentioned unambiguously associated with the action of microorganisms only have been found. Frequently, they could be a consequence of the action of not only and not so much microorganisms, but other factors also: temperature, light, aggressive media, mechanical loads, etc. Consequently, confirmation of the presence and quantitative estimation of microorganisms' participation in changing material properties requires model laboratory experiments on reproduction of type of the real microbiological damage.

Methodologically, such experiments are analogous to the above-considered ones, carried out at determination of microbial cells, detected in the real samples, belonging to microorganisms-destructors. They also provide for cultivation of microorganisms on the material with estimation of microbiological growth and changes of properties of the latter. Hence, in the model experiment, samples inspired by microorganisms, detected in the real samples, are incubated under conditions of the surrounding medium (temperature, humidity, etc.) close to real conditions of the article operation. Special samples considering construction features of damaged parts, aggregates and assemblies of the article can also be used.

Thus characteristics and methods shown in Table 3.1 can be used for determination of the presence and intensity of material interaction with microorganisms-destructors.

Abilities of suggested characteristics and methods to study quantitative regularities of these processes in model laboratory experiments are discussed in Chapters 4, 5 and 6.

Not all and not always separate characteristics and methods of their determination allow objective judging about proceeding of considered microbiological damaging during the study of operated machinery. This requires comparative analysis of separate determination of a series of parameters inherent to the current process (mentioned in Table 3.1) with the help of a selection of various physicochemical, biochemical and microbiological methods and reproduction of the process type in the model laboratory experiment. Such complex application of parameters and methods provides for reliable results of studied processes' determination obtained with their help and, consequently, allows reliable detection of microbiological damage of operated technical articles.

The scheme illustrating the technique of such investigation is shown in Figure 3.1. General conclusion about the presence and intensity of microbiological damaging the material must be based on the totality of all data obtained by analyses due to the scheme (Figure 3.1).

3.2. DAMAGES OF MATERIALS OF TECHNICAL ARTICLES INDUCED BY MICROORGANISMS

In accordance with the technique for carrying out investigations, shown in Figure 3.1, the damaging degree of machinery by microorganisms has been estimated under real conditions.

Investigation objects were technical articles of simple structure, production technology and application (protective and optical glasses, technical documents), and highly complicated ones (modern motor vehicles and aviation facilities).

Many cases of damages of various articles have been observed and studied. It has been found that the majority of materials used in technical articles are more or less subject to damaging effect of microorganisms.

Damages of materials displaying low microbiological resistance (wood, paper, fiber, leather, cotton fabrics, felt) give about 15% of the total quantity of detected damages. The reason of so low quantity of damages in these materials is that protection means and methods previously developed and used in the machinery significantly increase microbiological resistance of the appropriate articles under operation conditions. Simultaneously, analysis of circumstances of occurring detected damages testifies that they all are usually induced by non-compliance with the requirements of regulatory and technical documents on production and operation (storage) of such articles.

Polymeric (28.5%), combustible and lubricating (27.4%) materials, varnished covers (16.2%), metals and alloys (12.6%) are most often damaged by microorganisms. At the same time, in standard laboratory tests, the majority of these materials display significant microbiological resistance of satisfactory products. Consequently, the results obtained give an opportunity to suggest that the existing requirements and test methods of microbiological resistance reflect insignificantly the type of the microorganism-material interaction and influence of the actual conditions of technical operation on this process.

Table 3.2 shows species of microorganisms-destructors detected on damaged materials. Usually, samples picked from them possess several species of microorganisms related to different taxonomic groups. However, one or two species only are most often destructors of the materials. Among them, microscopic fungi and bacteria are the most widespread. The former have been detected in more than 80% of all studied samples, and the latter in 16% of cases. Along with these microorganisms, an insignificant quantity of ray fungi and

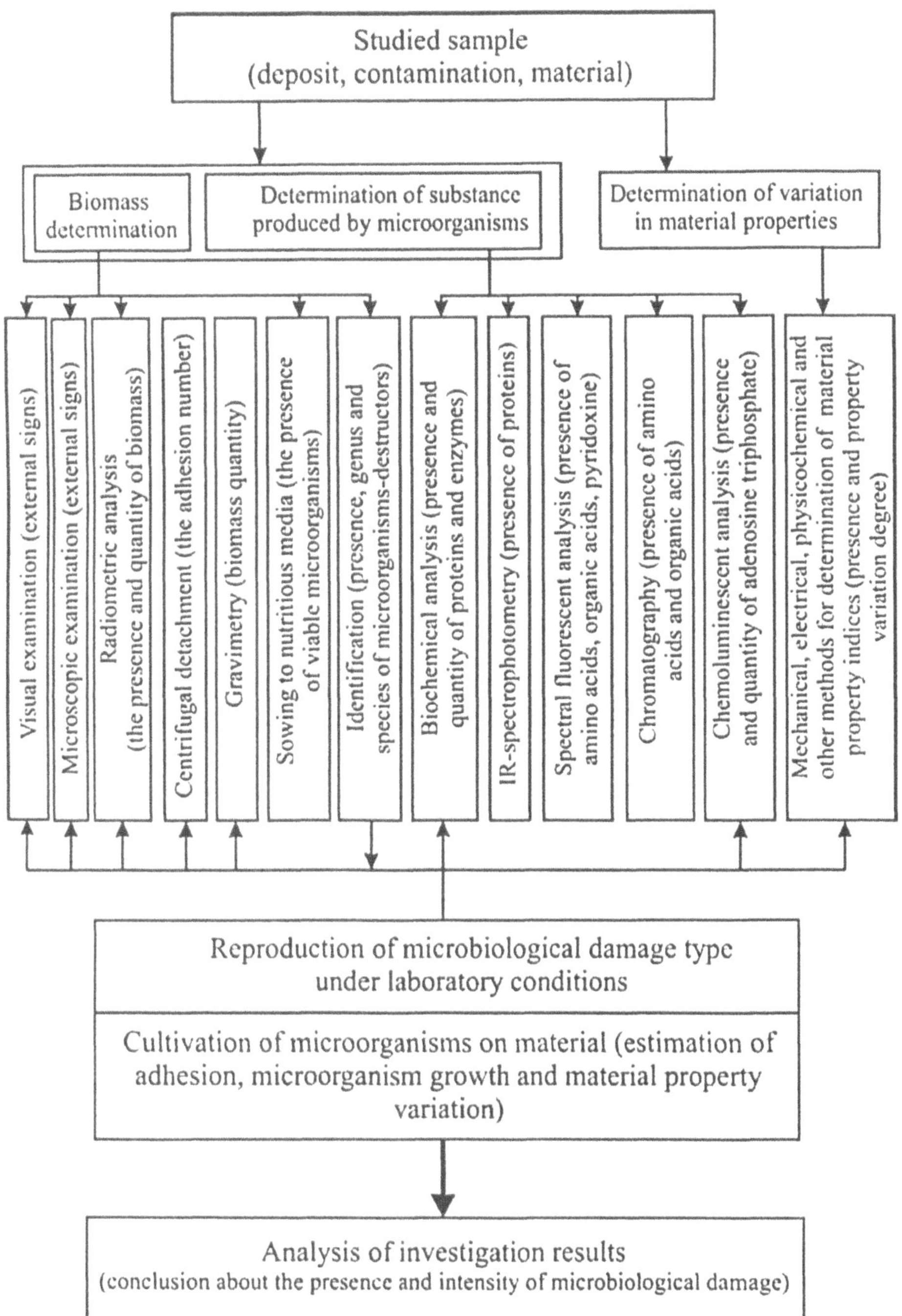

Figure 3.1. Scheme of detecting microbiological damaging of materials in operated technical objects

yeasts has been detected. However, they are significantly less aggressive to technical materials.

Fungi of Aspergillus, Penicillium, Trichoderma, Fusarium and Alternaria genus and bacteria of Bacillus genus are capable of damaging various types of materials. At the same time, many microorganisms possess the ability to destroy some special materials. Fuels are most often damaged by Cladosporium and Penicillium genus of fungi, as well as by Pseudomonas bacteria. More than a half of all cases of microbiological damaging of oils and lubricants are caused by the action of Aspergillus, Penicillium, Fusarium and Scopulariopsis genus fungi and Bacillus genus bacteria. Corrosion of metals and alloys induced by a microbiological factor is most often induced by Aspergillus, Penicillium, Alternaria and Paecilomyces genus fungi, as well as by Bacillus genus bacteria. In the majority of cases, microbiological damages of polymeric materials and varnishes is induced by colonization of these materials by Aspergillus, Penicillium, Stemphylium, Chaetomium, Trichoderma, Alternaria, Fusarium, and Bacillus genus. Optical glass is damaged by Aspergillus, Penicillium and Trichoderma genus.

The following species are the most aggressive in relation: to fuels – *Cladosporium resinae*, *Penicillium chrysogenum*, and *Pseudomonas aeruginosa*; to oils and lubricants – *Aspergillus niger*, *Penicillium chrysogenum*, *Fusarium sambucinum*, and *Bacillus subtilis*; to polymeric materials and varnishes – *Aspergillus niger*, *Aspergillus terreus*, *Aspergillus flavus*, *Penicillium chrysogenum*, *Penicillium funiculosum*, *Alternaria alternata*, *Stemphilium botryosum*, and *Bacillus subtilis*; to metals and alloys – *Aspergillus flavus*, *Penicillium chrysogenum*, *Penicillium funiculosum*, and *Bacillus subtilis*; to optical glass – *Aspergillus versicolor* and *Aspergillus flavus*.

Table 3.2 shows the main species of microorganisms damaging various types of materials.

Table 3.2

Microorganisms-destructors damaging technical materials

Materials	Species of microorganisms
Fuels	**Fungi**: *Cladosporium resinae*, *Cladosporium herbarum*, *Cladosporium chlorocephalum*, *Penicillium chrysogenum*, *Penicillium steckii*, *Penicillium notatum*, *Penicillium cyclopium*, *Penicillium nalgiovensis*, *Acremonium kilinse*, *Aspergillus niger*, *Furasium moniliform*, *Fusarium solani*, *Stemphylium botryosum*; **Bacteria**: *Pseudomonas aeruginosa*; **Yeasts**: *Hansenula holstii*.

Table 3.2 (continued)

Oils and lubricants	**Fungi**: *Aspergillus niger, Aspergillus caespitosus, Aspergillus flavus, Aspergillus tamari, Aspergillus fumigatus, Aspergillus melleus, Aspergillus ochracens, Penicillium chrysogenum, Penicillium rubrum, Penicillium verrucosum, Fusarium sambucinum, Fusarium solani, Scopulariopsis brevicaulis*; **Bacteria**: *Bacillus subtilis, Bacillus pumilus, Mycobacterium lacticolum.*
Materials based on synthetic and natural polymers	**Fungi**: *Aspergillus niger, Aspergillus flavus, Aspergillus terreus, Aspergillus amstelodomi, Penicillium funiculosum, Penicillium chrysogenum, Penicillium notatum, Penicillium cyclopium, Penicillium ochro-chloron, Alternaria alternata, Aureobasidium pullulans, Fusarium sp., Chaetomium globosum, Trichoderma viride, Cladosporium herbarum, Cladosporium sphaerospermum, Stemphylium botryosum, Paecilomyces varioti*; **Bacteria**: *Bacillus megaterium, Bacillus pumilus, Bacillus subtilis, Acinetobacter sp., Pseudomonas fluorescens, Mycobacterium globiforme*; **Yeasts**: *Hansenula holstii.*
Varnished covers	**Fungi**: *Aspergillus niger, Aspergillus flavus, Aspergillus terreus, Aspergillus amstelodomi, Aspergillus tamarii, Penicillium chrysogenum, Penicillium brevi-compactum, Penicillium funiculosum, Penicillium cyclopium, Penicillium ochro-chloron, Penicillium martensii, Trichoderma viride, Paecilomyces varioti, Alternaria alternate, Fusarium moniliform, Chaetomium globosum*; **Bacteria**: *Pseudomonas aeruginosa, Flavobacterium.*
Metals and alloys	**Fungi**: *Aspergillus niger, Aspergillus flavus, Penicillium chrysogenum, Penicillium funiculosum, Trichoderma viride, Aureobasidium pullulans, Cladosporium cladosporioides, Scopulariopsis bevicaulis, Fusarium sambucinum, Rhizopus nigricans, Alternaria sp., Paecilomycis varioti*; **Bacteria**: *Bacillus subtilis, Pseudomonas aeruginiosa.*
Glass (optical and protective)	**Fungi**: *Aspergillus niger, Aspergillus terreus, Aspergillus penicilloides, Aspergillus flavus, Aspergillus versicolor, Penicillium chrysogenum, Penicillium regulosum, Chaetomium globosum, Trichoderma viride, Paecilomyces varioti, Scopulariopsis brevicaulis.*

Analysis of investigation results on microorganisms-destructors has displayed that their species composition and properties of strains (aggressiveness to material, physiological, biochemical and antagonistic properties) differ due to the type (brand) of damaged material, construction and technological features of the parts (associates, aggregates), soil and climatic region of the article operation. These interconnections should be considered in selection of the test-cultures for testing microbiological resistance of materials and articles and the efficiency of protection means and methods.

The types of damages to studied materials induced by microorganisms, listed in Table 3.2, are shown in Table 3.3.

Table 3.3

The type and consequences of microbiological damaging of technical materials

Material	Type of damage to material	Consequences of damage affecting efficiency of article
Combustible materials and lubricants: aviation kerosene, diesel oil, and gasoline; lubricating and hydraulic oils; active and preservative lubricants	Fibrous slimy masses in the surface layer (lubricants) or in the volume of material, at the material-water interface (fuels, oils), stable emulsions (fuels, oils); color change; change in consistence: thinning or thickening (oils, lubricants); acidity variation; viscosity change (oils, lubricants)	Filter loading, breakdowns in operation of pumps and other units of fuel, oil and hydraulic systems of articles; destruction of protective coatings and hermetic sealers; corrosion, increase of friction and wearing of metal parts and assemblies (hinging and joining assemblies, joints, bearings, gearing mechanisms, reducers, hydraulic boosters, Shock-absorbers, etc.)
Materials derived from synthetic and natural polymers: organic glass, polyvinylchloride plasticate, polyethylene film, varnished fabric, glass-fiber laminate,	Blooms of different consistence and color; change of surface color and luster; change of mechanical, dielectric properties and compositions (reduction	Hermetic sealer swelling, inflation and exfoliation; organic glass cracking; electrical resistance reduction and destruction of wire insulation; short-

sailcloth, tarpaulin, rubberized fabric, heat and sound insulating materials, Capron, hermetic sealers, and rubbers	of plasticizer content, swelling, inflation, cracking)	circuiting and destruction of current-conducting traces of printed boards; change of electrical parameters of radio and electronic units (REU); reduction of strength characteristics and destruction of REU components; destruction of rubber bushing profiles and glands, etc.
Metals and alloys: aluminum alloys, steels, magnesium alloys, brass, copper alloys	Blooms of different form, color and consistence (slimy, powder-like, felt-like); metal corrosion (surface, pitting, stratifying)	Corrosion centers and destruction of load-bearing components of structures and fuel tanks from metal; corrosion of REU components; change of electrical parameters and breakdowns, etc.
Varnished coatings: primers, paints, lacquers, enamels	Blooms of different form, size and consistence; change of color and luster; inflation, cracking and exfoliation; changes in physicomechanical indices (strength, adhesion, etc.)	Destruction of covers and corrosion of metal parts of structures and equipment; destruction of covers and parts from polymeric materials (electrical insulation from varnished fabric, radio antennas, etc.); change of REU electrical parameters
Glass optical and protective	White or grey bloom spots of different shape and size; surface etching giving a thread-like pattern; light transmission variation (optical properties)	Deterioration of parameters of optical devices (lenses, color filters, binoculars, protective glasses, etc.)

For all materials, general unfavorable variation is formation of sediments (pollutions) representing products of microbial growth and metabolism (biomass, metabolites of microorganisms – organic acids, enzymes and other substances) on the surface (or in the volume).

The investigations performed confirm that the main reason of variations in the material properties, caused by microorganisms, is the action of metabolites. The authors of the present monograph have not observed cases of mechanical destruction of materials due to growth of fungal mycelium mentioned by many other investigators [2, 4].

All microbiological damages detected during operation of technical objects can be divided into two types due to the degree and features of metabolites' effect on the material properties. The first type of damages promotes reversible changes in the properties. This means that all detected changes in indices of the material are restored to the initial level after removing biological sediments. Many cases of reversible reduction of electrical resistance and strength properties of polyvinylchloride (PVC) and varnished fabric insulation of electric lines, glass fiber laminate printed circuit boards, reduction of strength of fiberglass based on poly(methyl methacrylate) (PMMA), polyethylene (PE) and cellulose triacetate (CTA) films, as well as the increase of content of mechanical admixtures in petroleum fuels, technical oils and lubricants have been observed for operated technical objects.

Data in the literature on the effect of liquid media on materials and analysis of experimental results reproducing microbiological damages, obtained in accordance with the scheme in Figure 3.1, give grounds to suggest that, above all, reversible changes in properties considered are the result of physical processes: sorptional interaction with metabolites for solids and chocking up by microbial masses for liquids.

The second type of microbiological damages is presented by the one causing irreversible changes in properties. This means that changes are observed both in the presence of microorganisms on the material and after removing them.

Irreversible changes of strength and dielectric indices associated with the effect of microorganisms have been observed for polyvinylchloride, varnished fabric and rubber insulation of electric wires; strength reduction was observed for rubbers of soft fuel tanks and hoses, cotton and synthetic fabrics.

Damages of this type are also cracking of fiberglass, degradation of varnished coatings, polysulfide hermetic sealers and glass fiber laminate, defects in metals and alloys, changes in acidity of combustible and lubricating materials and viscosity of technical oils and lubricants.

FACTORS PROMOTING OCCURRENCE AND DEVELOPMENT OF

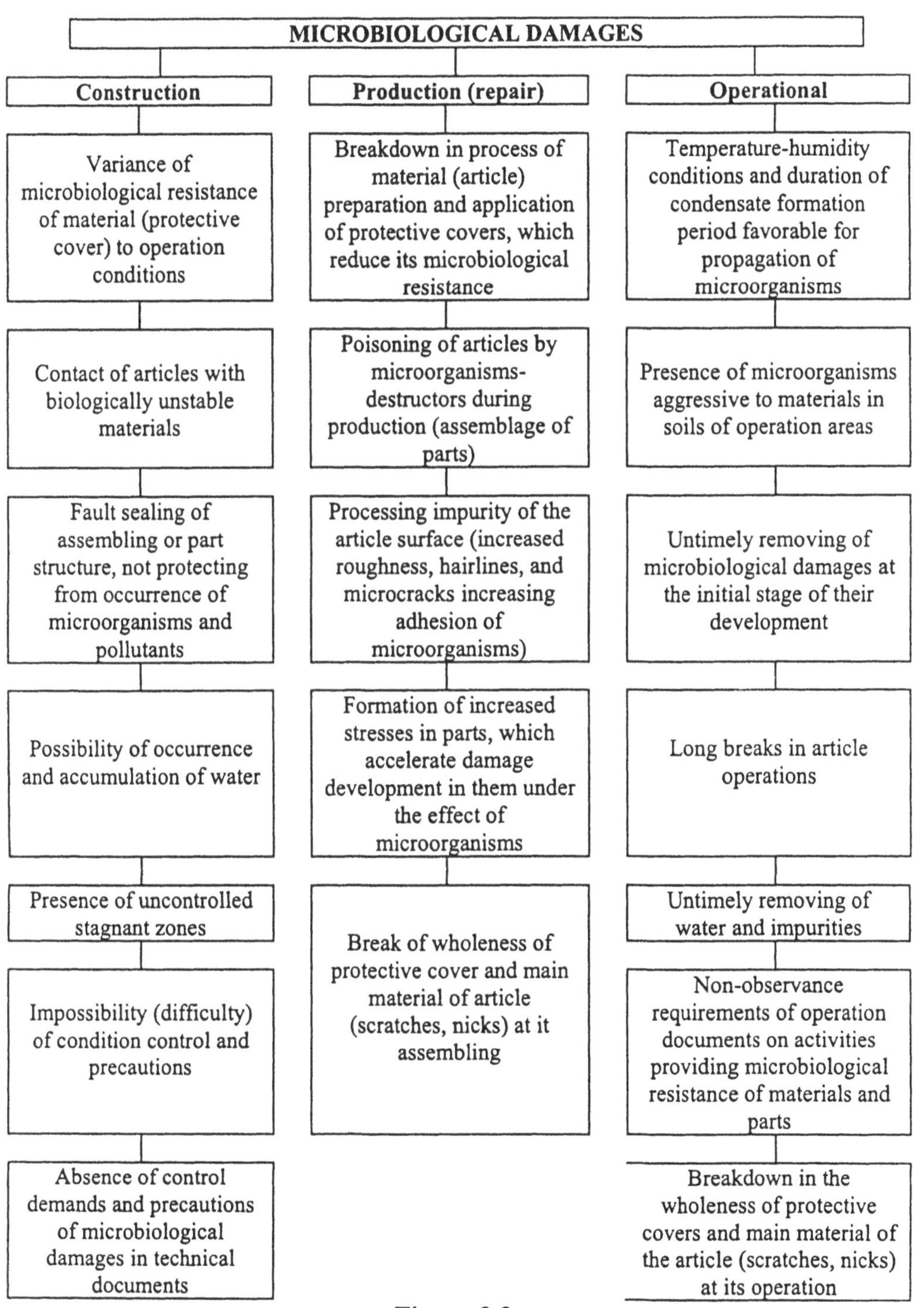
MICROBIOLOGICAL DAMAGES
Construction
Production (repair)
Operational
Variance of microbiological resistance of material (protective cover) to operation conditions
Breakdown in process of material (article) preparation and application of protective covers, which reduce its microbiological resistance
Temperature-humidity conditions and duration of condensate formation period favorable for propagation of microorganisms
Contact of articles with biologically unstable materials
Poisoning of articles by microorganisms-destructors during production (assemblage of parts)
Presence of microorganisms aggressive to materials in soils of operation areas
Fault sealing of assembling or part structure, not protecting from occurrence of microorganisms and pollutants
Processing impurity of the article surface (increased roughness, hairlines, and microcracks increasing adhesion of microorganisms)
Untimely removing of microbiological damages at the initial stage of their development
Possibility of occurrence and accumulation of water
Formation of increased stresses in parts, which accelerate damage development in them under the effect of microorganisms
Long breaks in article operations
Presence of uncontrolled stagnant zones
Break of wholeness of protective cover and main material of article (scratches, nicks) at it assembling
Untimely removing of water and impurities
Impossibility (difficulty) of condition control and precautions
Non-observance requirements of operation documents on activities providing microbiological resistance of materials and parts
Absence of control demands and precautions of microbiological damages in technical documents
Breakdown in the wholeness of protective covers and main material of the article (scratches, nicks) at its operation

Figure 3.2.

Model experiments on reproduction of the mentioned damages and present data on the influence of aggressive media on materials give an opportunity to suggest that they are induced by chemical processes: degradation of unstable chemical bonds (chemical degradation) for polymeric materials and varnishes and electrochemical corrosion for metals and alloys.

Irreversible changes of properties can also be caused by physical processes presumably residing in the sorption-diffusion interactions of stressed polymers with metabolites, as well as in diffusive desorption of low-molecular components from polymeric materials and varnish coatings.

Detailed description of model experiments forming the basis for the above-mentioned hypotheses about the type of process causing the influence of microorganisms on properties of materials is present in Chapter 6.

Analysis of observed cases of microbiological damages has displayed that their type and intensity of development depend on properties, conditions and features of the material application (in articles), aggressiveness of microorganisms-destructors, duration and conditions of interactions in the material-microorganism couple, as well as on some other factors promoting these interactions.

The most significant conditions for microbiological damaging are the temperature-humidity mode and the presence of mineral and organic pollutants in the material. The main factors promoting damages can be subdivided into three groups as shown in Figure 3.2. It has been found that every separate factor among these shown in Figure 3.2 promotes biological damage in one case or another. However, the determining effect on the possibility of their occurrence and intensity of development is usually caused by a combination of factors.

Sediments, material pollutions and variations of its properties occurring due to the action of microorganisms increase expenses for technical maintenance of the articles (cleaning, repair, substitutes, restoration of parts, etc.). At the same time, note that for technical facilities, impact on technical conditions and efficiency of parts, aggregates, units and systems of the article represent the most important index of microbiological damage.

Significance of the microbiological factor in occurrence of every particular case, generally described in Table 3.3, is determined with the help of specialists in the branch of operation of articles, separate systems and aggregates. Consequences discussed in the Table were induced by microbiological damages belonged to both types. For example, irreversible changes in properties and destruction of glass-fiber laminate, rubber and varnished coatings used in electrical and radio units frequently deteriorate efficiency of devices and equipment. Reversible microbiological damages of these materials do also induce frequent breakdowns, the so-called “floating effects”. After warming-up the device with damages of this kind, its electrical parameters are restored (in many cases, by even simple switching on the device).

Fuel and oil filter loading by microbial mass is the most typical consequence of microbiological damage of fuels and lubricants. The cases of fuel filter loading are frequently observed in aviation.

Studies of the breakdowns in operation of aircraft fuel filter have detected the reason which is the filter protective network loading by microbiological mass.

Fuel filter of ships are also frequently loaded by microbiological masses. Induced frequent substitution of filters is unfavorable for their efficiency.

It has been observed in laboratory experiments reproducing damages to fuel systems that development of microorganisms causes accumulation of organic acids in the fuel and aqueous deposit present in fuel tanks. The raise of the acid number from 0.35 to 1.2 – 1.3 mg KOH per 100 mg of fuel (after 180 days of growth) is observed. Hence, the aqueous phase pH is reduced from 7.0 to 4.9 – 4.3. Metabolites excreted by microorganisms induce inflation and exfoliation of hermetic sealers and varnished coatings, corrosion of metal parts applied in constructions of aggregates and units of the fuel systems of the articles. Studies of *Cladosporium resinae* fungus action (during 30 days) on samples of D-16 alloy with applied sealer U-30MES-5 have displayed strength reduction of the latter by 25 – 30%. Hence, corrosion (ulcers of continuous) of 100% of the sample surface is observed.

Fuel leakages from fuel tanks-boxes of aircrafts occurred during operation, caused by through corrosion of the walls induced by microbial growth in the fuel.

Biomass capable of inducing such consequences (damages) may occur in the fuel in the presence of free water, the fuel-water interface or mineral pollutants, as well as in the environmental temperature range of 10 - 30°C.

Complex studies of the reasons of microbiological damages indicate that they usually occur in case of combining various factors favorable for development of microorganisms (see Figure 3.2). For example, construction of transport airplane fuel tanks form conditions for water accumulation in undrainable fuel residues. Hence, control over its composition is practically impossible during operation until clear signs of the above-mentioned microbiological damages occur. Other factors promoting these damages are as follows: operation in tropical and subtropical climate and in areas with abrupt day and night temperature difference, which stipulate formation of water condensate in the tanks; long interruptions in flights; non-correspondence of microbiological resistance of materials used in construction of fuel systems (U-30MES-5 sealer and other protective coatings) with operation conditions; the absence of requirements for control over microbiological pollution degree of fuel systems in operational manuals for the cases of flight preparation and routine maintenance of airplanes, construction and operation conditions of which induce the above-mentioned damages.

Occurrence and development of corrosion of parts of oil of hydraulic systems (shock-absorbers, hydraulic cylinders, lifters, etc.) and greased parts of friction associates are quite frequent consequences of microbiological damage to oils and lubricants. In these cases, metabolites produced by the microorganisms significantly increase acidity of oils and lubricants raising their corrosion activity to metals.

Estimation of corrosion activity of oils and lubricants damaged by *Bacillus sp.* bacteria on St.3 steel samples has indicated corrosion ulcers depth from 0.19 to 0.79 mm during 180-day exposure for different materials. For control samples, it varied from 0.015 to 0.05 mm.

Other consequences of such damages to lubricants are increase friction stresses and wear off in joints, bearings, hinges, gearings, etc.

All these microbiological damages to oils and lubricants occur in the presence of free water in them (over 1 – 2%) or relative humidity over 70% and temperature from 15 to 30°C. They are also promoted by occurrence and accumulation of pollutants in these aggregates, as well as nonobservance of the time for lubrication substitution provided by operational documents.

The most serious microbiological damages to polymeric materials are caused by substances produced by microorganisms, which induce significant changes in mechanical, dielectric and physicochemical properties of materials up to destruction of articles and their components made from these materials. Decrease of electric insulation resistance and destruction of its external layer (varnished fabric coating) was observed many times in aviation.

Microbiological damages of polymeric materials, from which components of REU are made (components of printed boards, priming compounds, electric insulation lacquers, sockets and bearing brasses of plugs, etc.), induce changes in operational parameters and loss of efficiency of devices and equipment.

Model laboratory experiments with an electric circuit indicate that even a small adverse impact of microorganisms on one of its components made from polymers leads to a significant change of electric parameters of the whole circuit.

The impact of microorganisms on materials of parachute systems promotes reduction of their strength. Strength at break of the cords and the canopy is reduced by 15 – 20%.

Metabolites produced by microorganisms induce swelling and destruction of varnish coatings. Penetrating to the material protected by such coating, they promote development of metals' corrosion, change of structure, composition and properties of polymers and other materials.

Laboratory experiments have indicated that growth of *Aspergillus niger* microscopic fungus on some paints induce three-fold decrease of blow resistance already after 30 days of the impact. In the presence of a hydrophilic pollution, the fungus growth destructs the coating. At the same time, the control samples remained unchanged.

Microbiological damages to varnish coatings have frequently caused significant exfoliating corrosion of load-bearing components of structures from aluminum alloys, ulcer corrosion of components from magnesium alloys, and corrosion of parts from steel.

Microbiological damage to optical glass induced breaks in efficiency of photorecorders and navigation devices. The impact of metabolites of microorganisms reduces optical parameters of lenses, oculars, prisms, and color filters.

All the above-mentioned damages to parts and components of devices from various materials, induced by growth of microorganisms and their metabolites, may cause breakdowns in operation of units and serious and even lethal accidents in aviation, motor vehicles, ships, etc.

CONCLUSION

Suggested in the present Chapter is a complex of parameters and determination methods of the main stages of microbiological damaging of technical materials – adhesion, growth of microbial cells and changes in properties of materials impacted by microorganisms. Characteristics and methods give an opportunity to detect the presence and qualitatively estimate intensity of these processes in the model experiment and in studies of samples from materials under operation.

The method of detecting microbiological damages to operated materials and articles is experimentally proved. It is based on separate determination of a complex of parameters peculiar to the damage with the help of microbiological, physicochemical and biochemical methods. Application efficiency of the method to estimation of the microbiological ability to damage operated technical samples and studies of the reasons of their breakdowns and failure during operation is displayed.

The composition and trademarks of damaged materials and species composition of microorganisms-destructors are determined, as well as conditions and factors promoting occurrence and propagation of microbiological damages. The materials mentioned and microorganisms damaging them were used as the test facilities for future investigations of regularities at every stage of the microbiological damaging.

Chapter 4.

Adhesive interaction between microorganisms-destructors and materials

---

In accordance with the ideas about the mechanism of microbiological damage, developed by the authors of the present monograph, the first stage of the damage is adhesive interaction between microbial cells and the material surface.

The present Chapter displays investigation results of occurrence features and kinetics of fungus-destructor spore adhesion at their transfer to the material surface from the air. Analytical models and quantitative indices of this stage of microbiological damaging are suggested. The method of obtaining indices and the possibility of their application to estimation and forecasting of microbiological resistance of materials and efficiency of protection measures are proved.

4.1. QUANTITATIVE DESCRIPTION AND PARAMETERS OF THE PROCESS

As mentioned in Chapter 1, occurrence of the adhesive interaction with microorganisms is possible at transferring the latter from the surroundings to a quite thin zone at the material surface, where adhesion forces are effective. Spores of biodestructors were transferred to the material by precipitating from calm air to the samples located at different angles to the direction of spore motion (Chapter 2). Hence, the force of spore detachment from the surface (by which the adhesion force is estimated) was set by varying the sample tilt.

Figure 4.1 shows dependencies of the part of fungus spores (a) adhered by polymeric materials in the specific expense of spores (p) after the end of their precipitation from calm air at different tilts of samples (α). Evidently, the ratios between p and a obtained for all studied types of materials and species of microorganisms are close to the same linear dependence $a = f(p)$. This means that a is increased with the sample tilt to the direction of spores' motion from 20 to 80°.

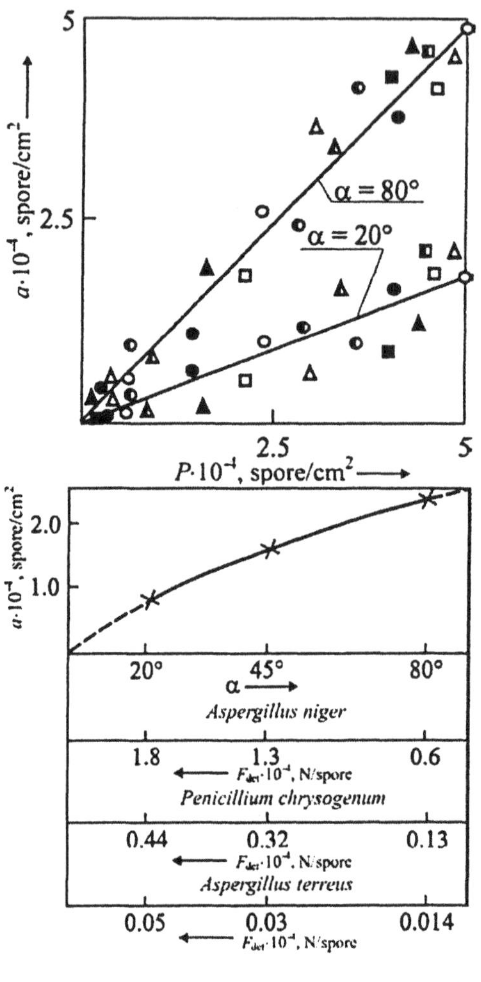

○ ● ◐ - *Aspergillus niger;* ○ □ △ - polyethylene;
□ ■ ◧ - *Penicillium chrisogenum;* ● ■ ▲ - cellophane;
△ ▲ ▲ - *Aspergillus terreus;* ◐ ◧ ▲ - poly(ethylene terephthalate)

Figure 4.1. Dependence of the quantity of spores adhered by polymeric materials (a): a - on the specific expense of spores (p) after the end of precipitation of spores; b – on the sample tilt (α) and detachment force (F_{det}) at specific spore expense of $2.5 \cdot 10^4$ spore/cm^2); α is the tilt of samples to the direction of spore precipitation.

Figure 4.1b shows also the values of F_{det} calculated[2] for various tilts of the surface and appropriate radii of spores of studied microscopic fungi. Analysis of these data shows that the quantity of adhered spores remains practically constant for various species of fungi and appropriate sizes of their spores, differing by values of their detachment forces. This means that the value of a is independent of F_{det} value used in the experiment. Hence, at precipitation from the air to polymeric surfaces, the quantity of adhered spores of fungi is independent of the species of microorganisms, type of material, but is determined by the specific expense of spores and the tilt of the sample.

Data obtained allow a supposition that the adhesion force of spores to the material is much higher than detachment forces used in experiments. In this case, all spores reaching the sample surface in the above conditions of their transfer must be kept on it.

Theoretically possible quantity of spores contacting with the sample has been calculated. When spores are precipitated to a plane surface located at tilt α to the direction of their motion, those shall be precipitated which occur within a hypothetic parallelepiped, the lower forming edge of which is the given surface. Then at the spore precipitation rate, v, and their concentration in the cloud, c, theoretically, the quantity of spores, a_T, shall appear on the specific (1 cm^2) surface during time t:

$$a_T = c \cdot t \cdot v \cdot \sin\alpha. \tag{4.1}$$

Substituting the multiplication $c \cdot t \cdot v$ by specific expense of spores, p, time of precipitation end is obtained:

$$a_T = p \cdot \sin\alpha. \tag{4.2}$$

Using experimentally obtained values of p, theoretically possible appropriate quantities of spores contacting with the sample surface can be calculated by equation (4.2).

It has been found that at the particular specific expenses of spores, values of a_T calculated by equation (4.2) and appropriate experimental values of a (Figure 4.1) are almost the same. Their deviation does not exceed 15 – 17%, which is close to the mistake of experimentally determined p and a.

---

[2] The detachment force was calculated from the expression: $F_{det} = 4/3\pi r^3 \rho g \sin(90 - \alpha)$, where r and ρ are the radius and the density of spores, respectively; g is the gravity constant.

As a consequence, the adhesion number (γ) equal to the ratio a/a_T (Chapters 1 and 2) is close to 1. This means that at studied detachment forces, fungus spores precipitated from the air are practically not removed from the polymeric surfaces.

Thus all spores of fungi transferred to the surface are adhered to it. Hence, independently of the type of material and the species of microscopic fungus, the adhesion force exceeding $10^{-12} - 10^{-11}$ N/sp occurs at the initial moment of contact and provides for retention of spores on the surface at their precipitation from calm air to the sample.

Figure 4.2 shows kinetic dependencies of the increase of microscopic fungus spores quantity adhered by polymeric materials during precipitation. Irregular increase of a value is observed. At the initial and the end stages of the process, changes in a value do not exceed 25 – 35% of its maximal level observed in the experiment. The main quantity of spores is adhered to the polymer surface during a short period of time from t_1 to t_2. Special durations of periods since the dispersion (t_d) till the end of precipitation of spores to the sample (t_e) are peculiar to every studied species of microscopic fungi (Figure 4.2).

To explain found kinetic dependencies and obtain appropriate analytical calculation models, common [267] ideas about precipitation of finely dispersed particles by gravity have been used. In accordance with these ideas, such particles are precipitated due to the gravity force only and, consequently, the precipitation rate and time of reaching the surface for a particle will be defined by its geometrical shape (radius) and density. In calm air, particles with the size within 1 – 100 μm are precipitated at a constant rate v_g, the value of which is determined from the Stokes law:

$$v_g = 2/9((\rho_T - \rho_{air})/\mu)\cdot g\cdot r^2, \quad (4.3)$$

where ρ_T is the particle density; ρ_{air} is the density of the medium where precipitation proceeds; g is the acceleration of gravity; μ is the medium viscosity coefficient; r is the radius of the particle.

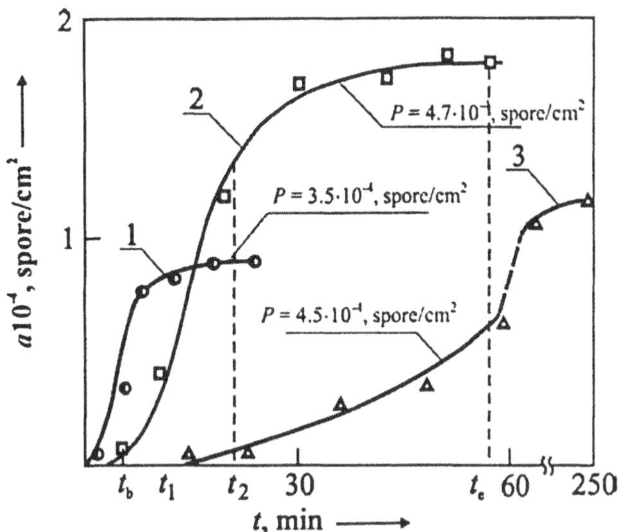

Figure 4.2. Dependence of the quantity of fungus spores (a) adhered by the material on time of sample exposure in the spore flow. (Tilt of samples is 20°; notations the same as in Figure 4.1)

As mentioned in Chapter 2, studied spores of various species of fungi possess almost the same density, but differ significantly in radius (by 1.5 – 4 times). Hence, significant dispersal in radii is also observed for the spores of every species of microscopic fungi. In this case, differences in radii can explain features of accumulation kinetics of fungus spores on polymers determined by the Stokes law (4.3).

If the gravitational approach to the studied process is true, expression (4.3) and obtained values of parameters of the spore distribution by radii (see Chapter 2) give a possibility to calculate precipitation rates, v_g^r, corresponded to different radii, r. Knowing the distances from the sample to the source of spores and rates, v_g^r, time of the beginning, t_b, and the end, t_e, of spore precipitation on the surface can be easily calculated.

Table 4.1 shows theoretical and experimental (Figure 4.2) values of t_b and t_e for spores of various species of microscopic fungi.

The data shown indicate satisfactory convergence of obtained results and experimental values of t_b and t_e. This confirms applicability of the notions and regularities of gravitational precipitation of finely dispersed particles of non-biological origin to precipitation of spores of the fungi-material destructors on the surface of polymers.

Thus the quantity of fungal spores adhered by the surface under conditions of their sedimentation in calm air is determined by the rate of precipitation (geometrical size of spores) and disposition of the sample surface in relation to the motion direction of spores.

Expressions (4.1), (4.2) and (4.3) give an opportunity to calculate parameters of the initial stage of adhesive interaction practically important for carrying out comparative and forecasting estimations: amount of spores adhered by the surface due to its disposition (distance from the source, tilt) and exposure in the spore flow, total duration, time of beginning and end of precipitation. The mentioned parameters can be determined for both spores of the same species and mixtures of spores of different fungus species.

To carry out appropriate calculations, one should know the experimental distribution of spores by radius and their specific expense or concentration in the air.

For comparative estimations, specific expense of the spores (or their concentration in the air) can be chosen from a quite broad range. For forecasting, it is desirable to use values of p (or c) close to really existing ones in the particular geographical region. Such values can be obtained as the result of specially executed investigations, as well as on the basis of data from the literature.

Table 4.1

Time of the beginning (t_b) and the end (t_e) of microscopic fungus spore precipitation to samples (the distance from the source of spores to samples is 70 cm)

Material	Fungus species	Maximal and minimal radius of spores, μm	Maximal and minimal rate of spore precipitation, cm/s	t_b^{exp} (t_b^{calc}), min	t_e^{exp} (t_e^{calc}), min
Poly(ethylene terephthalate)	*Aspergillus niger*	6.0 2.4	0.5 0.078	From 2 to 4 (2.3)	From 13 to 17 (15)
Polyethylene	*Penicillium chrysogenum*	4.0 1.2	0.22 0.02	From 2 to 5 (5.3)	From 50 to 55 (58)
Cellophane	*Aspergillus terreus*	2.5 0.6	0.085 0.005	From 10 to 15 (13.7)	From 230 to 260 (243)

Notes: t_b^{exp}, (t_b^{calc}), t_e^{exp}, (t_e^{calc}) are values of t_b and t_e, experimental and calculated by expression (4.2), respectively. Applied to calculations by expression (4.2) were values of the air density and viscosity corresponded to temperature-humidity conditions of the tests: ρ_{air} = $1.27 \cdot 10^{-3}$ g/cm^3, μ = $1.8 \cdot 10^{-4}$ g/cm·s. Note that viscosity depends insignificantly on changes in temperature and humidity, and the air density can be neglected, because it is much lower than the density of spores.

Note that the present quantitative methods of determining concentration of microorganisms in air are not adapted well to the studies of microscopic fungi and bacteria-destructors under natural conditions. That is why there are short data on concentrations of such bioagents in the atmosphere. At the same time, the Koch method is recommended in GOST 9.053[3] [246] and widely applied. It provides for precipitation of microorganisms from the atmospheric air to nutritious agar media, i.e. for determination of the specific expense of spores. There is also enough volume of information in special literature, obtained with the help of the mentioned method on seasonal variations of the quantity and species composition of microorganisms-destructors in various climatic regions. These data can be used for making forecasting calculations by formulae (4.1) – (4.3).

As mentioned above, sedimentation in calm air is the simplest ("ideal") mechanism of the fungus spore transfer to the material surface. This process is much more complicated under natural conditions. That is why, in addition, adhesion of *Aspergillus niger* spores to polymeric materials from turbulent air flow oncoming to the sample has been studied. Tests were executed in the same chamber, used for precipitation of spores in calm air (Chapter 2). A ventilator was mounted in a hole in the chamber bottom. It provided for vertical air flow and spores rising at the given initial velocity. The source of spores (discs from cellophane with a spore-containing biomass) was placed at the bottom, and samples – in the upper part of the chamber at 20° tilt to direction of the ascending air flow.

The samples were exposed in the spore flow during 10 min. During exposure, air was regularly sampled from the chamber for determining the average (in the current experiment) concentration of spores. Samples were taken via special holes in the chamber wall by a syringe with some water in it. Spores contained in the air sample were suspended in this water and their quantity was calculated with the help of Goryaev's chamber.

Basing on statistical information about the wind speed in different climatic zones [247], the air flow speed chosen for the tests was close to the one really existing under natural conditions and equal ~2 m/s. Ventilator promptness (n_v) necessary for forming such speed was calculated using the initial data shown in Table 4.2 and values and dimensionalities of parameters, assumed for aerodynamic calculations [248, 249].

---

[3] Russian state standard.

Table 4.2

Initial data for calculation of the ventilator promptness (n_v) at the air flow speed of ~2 m/s

Parameter	Designation, dimensionality	Value
Spore weight	G, kg	10^{-11}
Spore diameter	D, m	10^{-5}
Square of spore middle cross-section	$S = (\pi D^2)/4$, m^2	$78.5 \cdot 10^{-12}$
Diameter of the hole for air input	D_0, m	0.03
Square of the hole	F_0, m^2	0.0007
Ventilator propeller diameter	D_v, m	0.11
Velocity of sound	a, m/s	340
Meteorological conditions (common atmosphere):		
Air temperature	T, K	288
Air pressure	p, kg/m^2	1.0320
Air density	ρ, $kg \cdot s^2/m^4$	0.125
Air constant	m, $deg^{0.5} \cdot s^{-1}$	0.4

Values of n_v were calculated as follows. The Mach number (M) is determined from the expression:

$$M = v/a.$$

At v = 2 m/s, the Mach number equals 0.0059. The values of reduced specific air consumption $q(M)$ equal 0.01 was determined from the tables of gas dynamic values [250] by the value of M obtained. Then the specific air consumption, G_{air}, equals:

$$G_{air} = \frac{q(M) \cdot m \cdot F_0 \cdot p}{\sqrt{T}} = 0.0017 \text{ kg/s}.$$

Ventilator with the propeller of 0.11-m diameter creates such specific consumption at the rate of 2.5 rot/s.

After 10-min exposure in the air flow of spores, the sample materials were removed from the chamber, tested on a microscope, and microphotographs of them were taken. The microphotographs were used for counting spores adhered by samples.

It has been found that spores are adhered by both sides of the sample (front and back to direction of the air flow with spores), and on side edges, too. Hence, the quantity of spores at the edges (of both front and back sides of the sample) is denser than at the middle part of the samples. The results of visual inspection indicate that selected test mode reflects interaction of several mechanisms of the spore transport to the surface. Total quantity of spores adhered to the material will supposedly determined by combined proceeding of such processes as inert precipitation, edge effect of airflow, and sedimentation.

It has been found that the quantity of spores (a) on the front surface of the studied samples from poly(methyl methacrylate), glass fiber laminate and varnished fabric towards direction of the air flow are almost the same and fall within the range of $1.2 - 1.5 \cdot 10^5$ spore/cm^2. One may conclude that in both cases of sample airflow and calm air, the material type causes no significant effect on the quantity of fungus spores adhered to the surface.

Experimentally determined concentration of spores in the chamber, average during the experiment, equals 180 spore/cm^3. At such concentration, the value of a calculated by equation (4.1) at $v = 2$ m/s and $t = 600$ s is $6.4 \cdot 10^6$ spore/cm^2, e.g. by 1 – 1.5 decimal degrees greater than the value obtained in the experiment.

As mentioned above, equation (4.1) is true for sedimentation of the spores to the material surface from calm air. Obviously, it does not reflect features of the transfer mechanism, realized in the actual test mode. Another reason of differences between theoretical and experimental values of a can be reduction of the quantity of adhered spores due to the influence of the front force on them (the front resistance force) formed by the air flow. This force can exceed the adhesion force.

The value of the front resistance of a fungus spore has been determined:

$$Q = c_X \cdot q \cdot S_r,$$

where c_X is the coefficient of the front resistance of "adhered" spore equal 0.01; $q = 0.5 \cdot \rho \cdot v^2$ is the flow strength. The value of Q calculated by the above-mentioned expression equals $2.2 \cdot 10^{-12}$ N/spore. The value obtained is close to acting forces, formed at sedimentation of *Aspergillus niger* spores on the surfaces disposed at different tilts to the flow direction ($0.6 \cdot 10^{-12} - 1.8 \cdot 10^{-12}$ N/spore, Figure 4.1). As shown above, the effect of such forces causes no removal of the spores from polymeric materials.

Thus at the initial stage of adhesion, the quantity of spores adhered to the surface will mostly depend on the mechanisms of their transfer from the air to the zone of direct action of adhesive forces.

Figure 4.3 shows experimental integral curves of adhesion, obtained after some time of contact between the spores and the surface: dependencies of the adhesion number (γ_t) of fungus spores to polymeric materials on the affecting detachment force intensity (F), formed by the sample centrifugation method[4].

Curves of $\gamma_t = f(F)$ function testify about the present difference between adhesion forces of spores of the current microorganism species to the surface and properly characterize distribution of the spores by forces of their adhesion to this surface. Every interval on the integral curve corresponds to the part of spores (γ_1 - γ_2), the value of the adhesion force of which falls within the range of (F_2 - F_1).

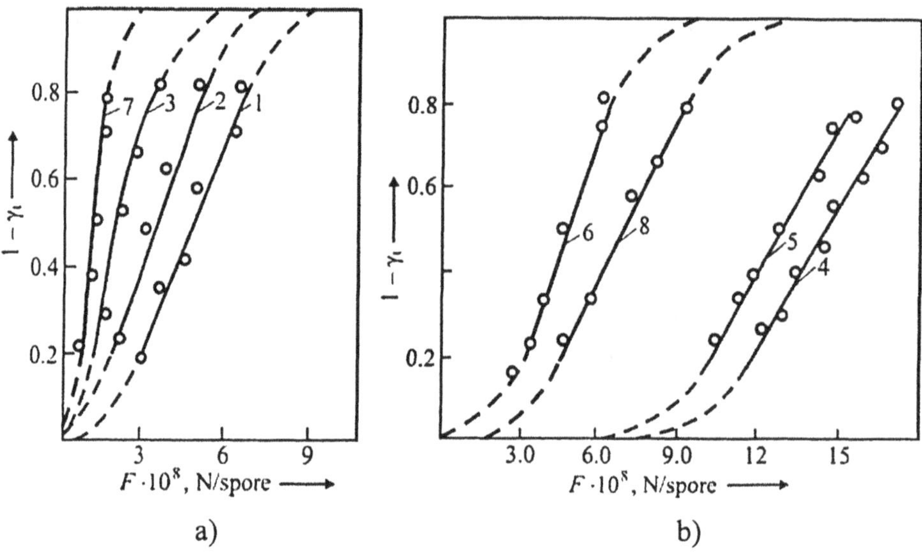

Figure 4.3. Integral curves of *Aspergillus niger* (curves 1 – 6) and *Aspergillus flavus* (curves 7 and 8) spore adhesion to polyethylene (a) and cellophane (b) after different time of contact with the materials Curves 3, 6, and 7 – 1 hour; curves 2 and 5 – 4 hours; curves 1 and 8 – 12 and 24 hours; curve 4 – 6 and 12 hours. Temperature – 29°C, humidity – 98%.

All curves shown in Figure 4.3 are of the same shape that displays the unity of regularities defining distribution of fungus spores by forces of adhesion to various materials. It is also observed that at increase of the spore contact time with surfaces, appropriate integral curves are shifted by the abscissa axis

---

[4] The value of acting detachment force was calculated by the formula from ref. [187]: $F_{det} = 4/3\pi r3\rho(r\pi n/60)R$, where r and ρ are the radius and the density of the spores, respectively; n is the number of centrifuge rotor rotations; R is the distance from the rotation axis to the surface polluted by the spores (the centrifuge constant).

towards the increase of adhesion (detachment) forces. E.g. adhesive interaction is of the kinetic type – the adhesive forces are formed not immediately, but during a noticeable time.

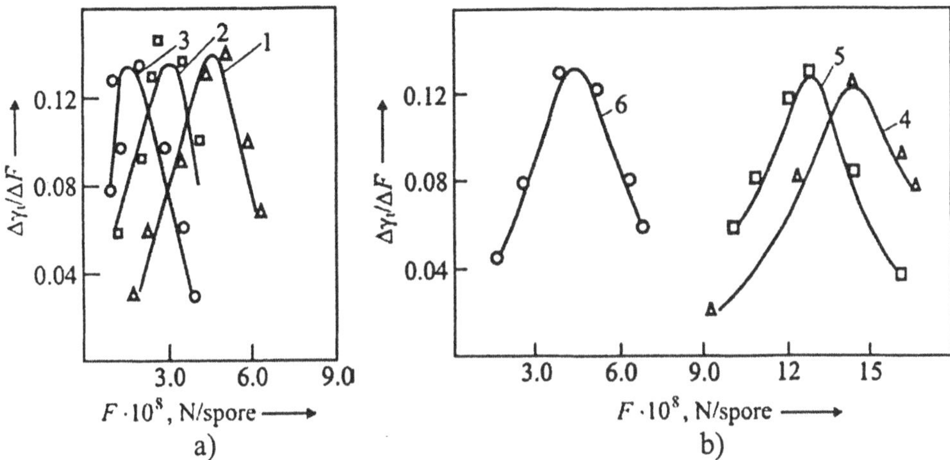

Figure 4.4. Envelopes of histograms of the adhesion number (γ_t) determination at various detachment forces (F) for adhesive interaction between *Aspergillus niger* spores and polyethylene (a) and cellophane (b). Duration, conditions of the spore contact with the surface and curve designations are the same as in Figure 4.3.

The detected distribution of spores by adhesion forces can be explained by inhomogeneity of their sizes, energetic and other properties of contacting surfaces, inconstancy of contact conditions, mistakes in performance of appropriate measurements, and other inconsiderable factors. This stipulates the statistical random type of the adhesion force and possibility to apply methods of mathematical statistics to analysis of $\gamma_t = f(F)$ dependence. In this case, the problem is reduced to detection of the frequency function $f(x)$ of the value $dN/N_0(F)$, where dN is the quantity of the spores possessing the adhesion force F; N_0 is the total quantity of the spores contacting with the material surface.

Distribution of the spores by adhesion forces $f(x)$ is identical to the distribution $f(F)$ of the value characterizing the adhesion force to the material surface of all spores present on it. E.g. $f(x) = f(F)$.

Then:

$$dN/N_0(F) = f(F).$$

Plurality of the factors defining formation of the adhesion force allows a supposition that their totality may lead to a distribution of the spores by adhesion forces according to the normal Gauss law.

This hypothesis is confirmed by the envelope of histograms $(\Delta\gamma_t/\Delta F) - F_1$, compiled on the basis of experimental dependencies $\gamma_t = f(F)$. Clearly, all envelopes possess the bell shape, typical of the normal distribution. For more strict proof of this hypothesis, the correspondence between experimental, theoretical normal, exponential and gamma-distributions was checked by the χ^2 criterion.

It has been found that for studied adhesive couples, the value of $P(\chi^2)$ for the normal law falls within the range of 0.82 – 0.98. For exponential and gamma-distributions, it has not exceeded 0.59. Hence, by forces of adhesion to the material surface, the spores of microscopic fungi are distributed according to the normal Gauss law:

$$\frac{dN}{N_0(F)} = \left(\frac{1}{\sigma_t^F\sqrt{2\pi}}\right)\exp\left(-\frac{F - F_t^{50}}{2\left(\sigma_t^F\right)^2}\right),$$

where F_t^{50} and σ_t^F are distribution parameters: mathematical expectation (mean value) and mean square deviation, respectively.

According to the symmetry property of the normal distribution, the mathematical expectation (mean value) of the adhesion force F_t^{50} considers adhesion forces of all the spores present on the material. The value of this parameter equals the force, at which 50% of all spores of fungi adhered to the surface will be removed from it, i.e. $\gamma_t = 0.5$. The higher F_t^{50} is, the higher is the force adhering spores of a microorganism to the materials.

The mean square deviation characterizes dispersion of the spores by adhesion forces in relation to the part of them, the adhesion force of which equals F_t^{50}. To put it differently, the value of σ_t^F indicates the difference between the minimal force field intensity, the effect of which causes removal of at least insignificant quantity of spores and that maximal detachment force, which should cause almost complete cleaning of the surface from the spores.

The quantity of spores possessing an adhesion force different from F_t^{50} increases with σ_t^F, in particular, the part of them, removing of which demands force field intensity, exceeding F_t^{50}. The higher σ_t^F is, the more spores are adhered to the material by the force exceeding F_t^{50} and, consequently, all other factors being the same, such material is the more so inclined to adhesive interaction with a biodestructor.

Values of the normal distribution parameters, shown in Table 4.3, indicate that the type of the material and species of the microorganism are sufficient for the F_t^{50} value. Each of studied adhesive couples is characterized by the unique value of this force. Consequently, one may suggest that the parameter F_t^{50} is sensitive to the features of the adhesive interaction between the microorganism and the material and perspective for application as an objective quantitative index of this process.

The results in Table 4.3 indicate also that the mean square deviations, σ_t^F, are almost the same for all adhesive couples. Therewith, correlation between F_t^{50} and appropriate mean square deviations is not observed.

Table 4.3

Distribution parameters for *Aspergillus niger* and *Aspergillus flavus* spores by intensities of the adhesion force to polyethylene and cellophane.

Temperature – 29°C, air humidity – 98%, contact duration between spores and material – 24 hours.

Material	Species of microscopic fungus	$F_t^{50}\cdot 10^8$, N/spore	$\sigma_t^F \cdot 10^8$, N/spore
Polyethylene	*Aspergillus niger*	5.1	2.8
	Aspergillus flavus	2.5	2.3
Cellophane	*Aspergillus niger*	14.2	2.7
	Aspergillus flavus	6.2	2.5

Thus a single application of parameter F_t^{50} (hereinafter, the average adhesion force) or combined application of parameters F_t^{50} and σ_t^F, which allow complete characterization of the spore distribution by forces of their adhesion to various surfaces, is enough for quantitative estimation of the adhesive interaction.

To clear out reasons causing detected distribution of the fungus spores by forces of adhesion to the polymeric surface, model experiments excluding or reducing to the minimum the effect of inhomogeneity in geometrical sizes of spores on this distribution have been carried out. Monodisperse spherical particles from polystyrene 2.8 μm in diameter, produced by SERVA Co. (Germany), were used. The contact between particles and the surface of polymeric materials and their detachment were executed as in appropriate studies of adhesion of microscopic fungus spores. It was found that the type of integral curves of polystyrene particles adhesion is analogous to that of fungus spores. Consequently, the principal picture of adhesion of these two finely dispersed objects to the polymeric surface is the same. Therewith, it is known [112] that the existence of polystyrene particles distribution by adhesion forces is mostly associated with energetic inhomogeneity of contacting surfaces.

Calculations have indicated that distribution of polymeric particles by adhesion forces fit the normal law. The parameter F_t^{50} equals $9.2 \cdot 10^{-8}$ N/particle which is comparable with values of the analogous parameter for spores of microscopic fungi (Table 4.3). Simultaneously, σ_t^F equals $0.4 \cdot 10^{-8}$ N/particle, which is significantly, by 5 – 6 times lower than the mean square deviations obtained during the study of microscopic fungus adhesion. At the same time, analysis of data from Tables 4.3 and 2.2 (Chapter 2) shows that the values of σ_t^F are close to the appropriate mean square deviations of the fungus spore distribution by radii (Table 2.2). This induces a conclusion that inhomogeneity of the spore size is the dominant factor promoting their distribution by adhesion forces, at least, for the studied group of microorganisms. Suggested indices σ_t^F and F_t^{50} characterize conditions of the adhesive interaction at a definite moment of contact between the spores of fungi and surfaces of materials. At the same time, they do not reflect a kinetic aspect causing formation of the adhesion force.

Figures 4.5 and 4.6 show kinetic curves of changes in the adhesion number (γ_t) distribution parameters by adhesion forces of *Aspergillus niger* and *Aspergillus flavus* spores to polyethylene and cellophane. Clearly all graphical dependencies $\gamma_t = f(t)$ and $F_t^{50} = f(t)$ possess similar shape, close to the exponent, which allows a suggestion that for different microorganism-material couples, an increase of the adhesion force with time is general and kinetically regulated. Hence, the dependencies shown in Figure 4.5 indicate that every microorganism-material couple is characterized by its own time of adhesive force formation, i.e. the time of kinetic curve entering the plateau and maximal values of F_t^{50} and γ_t corresponded to this plateau.

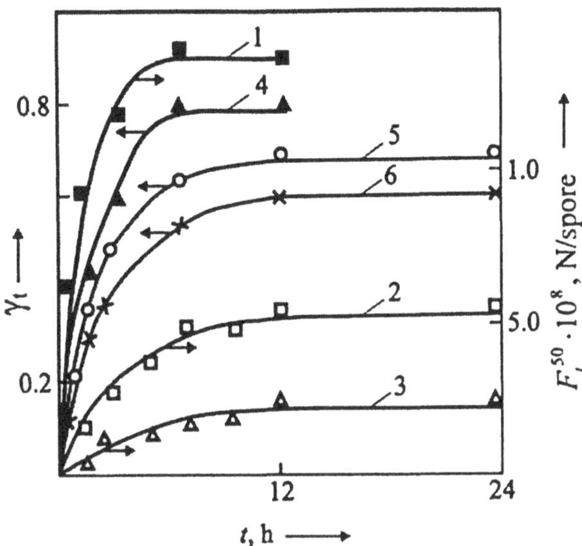

Figure 4.5. Kinetic dependencies of parameters F_t^{50} (curves 1, 2 and 3) and γ_t (curves 4, 5 and 6) for adhesive interaction of *Aspergillus niger* (1, 2, 4, 5, 6) and *Aspergillus flavus* (3) spores with polyethylene (2, 3, 5, 6) and cellophane (1, 4).
Values of γ_t have been obtained under the action of different intensities of detachment force: curve 4 – $12.0\cdot 10^{-8}$ N/spore; curve 5 – $3.7\cdot 10^{-8}$ N/spore; curve 6 – $4.5\cdot 10^{-8}$ N/spore. Temperature – 29°C; air humidity – 98%.

Kinetics of σ_t^F variation (Figure 4.6) display different features. For all studied adhesive couples, σ_t^F increases during 1 to 2 hours of contact, and then remains constant during the rest time of the adhesive interaction. One may suggest that such a type of σ_t^F variation is provided by processes of setting some primary equilibrium state between the material surface and fungal spores, for example, by setting the balance of water content in these two interacting objects. Water absorption or drying of spores, in this case, induces some initial variation of their geometrical sizes, which as shown above play the determining role in distribution of spores by forces of adhesion to the material. After setting such an "equilibrium" state in the system, σ_t^F is stabilized remaining unchanged during the rest time of the test. Hence, the parameter σ_t^F gives incomplete description to the kinetic features of the adhesive interaction between fungal spores and the surfaces.

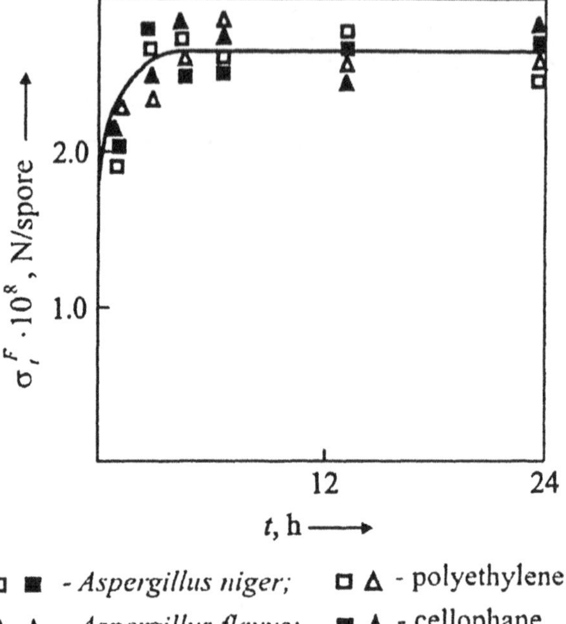

Figure 4.6. Kinetic dependencies of parameter σ_t^F of adhesive interaction of *Aspergillus niger* and *Aspergillus flavus* spores with polyethylene and cellophane.

The analysis of experimental data shown in Figure 4.5 has indicated that kinetic curves of increase of the average adhesion force, F_t^{50}, and the adhesion number, γ_t, is approximated satisfactorily by the general equation of the following type:

$$X_t = X_\infty[1 - \exp(-K \cdot t)], \tag{4.4}$$

where X_t is the value of F_t^{50} or γ_t at the moment of time t; X_∞ is the border (maximal) value of F_t^{50} or γ_t reached during the experiment; K is the rate constant of the process.

Maximal relative mistake of the approximation of experimentally obtained F_t^{50} or γ_t values from equation (4.4) equals 12 – 18% with regard to the type of material and the species of microorganism. Values of F_∞^{50}, γ_∞ and K obtained for the mentioned equation are shown in Table 4.4.

The value F_{∞}^{50} reflects distribution of spores by forces of adhesion after the completion of their adhering to the material surface. Constant K characterizes kinetic behavior of the system and represents the rate constant of adhesive force occurrence. Besides the maximal average force of adhesion, F_{∞}^{50}, constant K is desirable to be used as the index of adhesive interaction between the spores of fungi and the surfaces.

Constant K was calculated by equation (4.4) using preliminarily calculated values of F_{t}^{50}. Moreover, the value of K was also determined by another method – by equation (4.4) using primary experimental information about kinetics of γ_t variation (at the constant detachment forces).

Values of K obtained using these two groups of data are shown in Table 4.4. It has been found that the difference between appropriate values of the rate constant of adhesion does not exceed 10 – 12%. As a consequence, this kinetic index of the adhesive interaction can be determined directly from the experimental kinetic dependencies of the adhesion number. This conclusion is illustrated by Figure 4.7. Clearly all obtained values of F_{t}^{50} and γ_t (for the particular microorganism-material system), shown in coordinates of linearized equation (4.4), are approximated well by a single linear dependence, tangent of which equals the constant K.

Note that when the run of such kinetic curve is determined with satisfactory accuracy (at some detachment force) and using equation (4.4), one can calculate values of both K and F_{∞}^{50}. However, such calculation method is possible only if all the values of γ_t fall within the range of 0.2 – 0.8 during the whole kinetic experiment. As mentioned in Chapter 2, the reliability of γ_t values out of this range is abruptly reduced that may induce significant mistakes in determination of constant K and F_{∞}^{50}.

To determine the reasons of observed time dependence of the adhesion force, microscope tests of spores were carried out during their adhesive interaction with the surface. The shape and size of spores were determined with the help of electron scanning microscope OPTON DCM 950. The spores were fixed according to the methodology described in ref. [128] on gold plating.

Table 4.4

Parameters of equation (4.4) for adhesive interaction between microscopic fungi and polyethylene and cellophane. Temperature – 29°C, humidity – 98%.

Material	Species of microscopic fungus	$F_{\infty}^{50}\cdot 10^8$, N/spore	K_F, hour^{-1}	γ_{∞}, % (F_{det}, N/spore)	K_{γ}, hour^{-1}
Polyethylene	*Aspergillus niger*	5.1	0.27	0.72 ($3.7\cdot 10^{-8}$)	0.31
				0.61 ($4.5\cdot 10^{-8}$)	0.29
	Aspergillus flavus	2.5	0.32	0.68 ($1.7\cdot 10^{-8}$)	0.36
Cellophane	*Aspergillus niger*	14.2	0.46	0.78 ($12\cdot 10^{-8}$)	0.48
				0.6 ($13.5\cdot 10^{-8}$)	0.50
	Aspergillus flavus	6.2	0.53	0.8 ($1.7\cdot 10^{-8}$)	0.47

Note: K_F and K_{γ} are values of the rate constant K of the adhesive force formation, calculated from dependencies $F_t^{50} = f(t)$ and $\gamma_t = f(t)$, respectively.

It has been found that the spores of fungi are deformed during adhesion. Hence, their cross-sections parallel to the material surface increases by 100 – 150% during 24 hours of contact. Obviously, the contact square with the surface is also increased that, in its turn, causes raise of the adhesion force.

Hence, obviously, data of kinetic investigations show that the change of the adhesion force with time is associated with the increase of the adhesion contact square between the spores of fungi and the surface. Kinetics of the change in conditions of adhesive couple is described by an exponential equation. The exponential parameter of this equation represents the rate constant of the adhesive force formation.

The results obtained allow a conclusion that the following indices are desirable to use for full characterization of the adhesion features of microscopic fungus spores to the surfaces: mathematical expectation (mean value) of the maximal adhesion force F_{∞}^{50} (the average adhesion force), the mean square deviation of this value characterizing distribution of the spores by adhesion forces, as well as the rate constant of formation of these forces, K. Indices of the adhesive interaction are obtained on the basis of a series of experimental dependencies of the adhesion number on the detachment force at different time intervals of contact between fungus spores and the material and consequent calculation of the values of F_{∞}^{50}, (σ_{∞}^{F}), and K using the whole massif of experimental data.

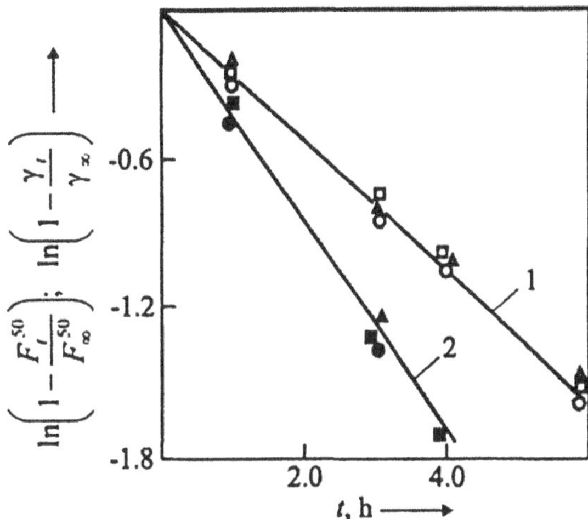

Figure 4.7. Linearized kinetic dependencies of parameters F_t^{50} (■, □) and γ_t of *Aspergillus niger* spore adhesion to polyethylene (straight line 1) and cellophane (straight line 2). The value of γ_t was determined for the detachment forces as follows: $3.7 \cdot 10^{-8}$ N/spore (○); $4.5 \cdot 10^{-8}$ N/spore (Δ); $13.5 \cdot 10^{-8}$ N/spore (•); $12.0 \cdot 10^{-8}$ N/spore (▲). Temperature – 29°C, humidity – 98%.

4.2. DEPENDENCE OF ADHESION ON PROPERTIES OF MATERIAL, SPORES OF MICROSCOPIC FUNGI AND TEMPERATURE-HUMIDITY CONDITIONS OF THE SURROUNDINGS

Possibilities of applying indices F_∞^{50} and K and methods of their determination to obtaining comparative estimations of the adhesive interaction between spores of fungi and different surfaces, the methods of their control and forecasting have been studied. Functional convenience of the suggested indices for solving the mentioned tasks has been confirmed by existence of the regularities of interconnection between F_∞^{50} and K and appropriate property indices of interacting materials and spores of fungi, as well as parameters of the environment.

Adhesion of finely dispersed particles of non-biological nature (dust, powders), as well as bacterial cells to the surfaces is usually associated with the surface properties of interacting objects, their hydrophobic-hydrophilic balance and roughness (irregularity), and size (radius) of finely dispersed particles [251 – 256]. These parameters of polymeric materials and spores of microscopic fungi were used in the analysis.

Table 4.5

Indices of the adhesive interaction between *Aspergillus niger* spores and polymeric materials. Temperature - 29°C, humidity – 98%

Material (notation)	Standard water sorption Q, %	$F_{\infty}^{50}\cdot10^8$, N/spore	K, hour^{-1}
Polyethylene (PE)	0.03	5.1	0.27
Epoxy resin (ER)	0.05	6.7	0.31
Poly(methyl methacrylate) (PMMA)	0.4	8.5	0.38
Poly(ethylene terephthalate) (PET)	0.5	9.3	0.39
Glass fiber laminate (GFL)	0.8	10.0	0.40
Varnished fabric (VF)	2.0	11.1	0.40
Cellulose triacetate (CTA)	3.5	12.3	0.43
Cellophane (CP)	30.0	14.2	0.46

Table 4.5 shows the adhesion indices of *Aspergillus niger* spores to various materials. Clearly, the fungus spores adhere to all studied polymeric surfaces. Every adhesive couple is characterized by the self selection of F_{∞}^{50} and K values. Comparative estimation, carried out using these indices, has allowed disposition of polymeric materials in Table 4.5 in the ascending order of magnitude of the adhesion force of *Aspergillus niger* spores to them. Therewith, for the series of materials shown in the Table, indices F_{∞}^{50} and K change symbately.

Hydrophobic-hydrophilic balance of the surfaces was characterized by the edge (contact) wetting angle (θ) of spores and materials by water [267]. The wetting angle, θ, of spores was determined on their dense layer formed on a filter after filtering water suspension of spores through it. To determine the contact angle of polymeric materials, samples preliminarily washed from pollutants were used. A drop of water 1 – 1.5 mm in diameter was applied to the surface by a syringe and θ value was measured by horizontal microscope MG-1.

It is found that the edge wetting angle of *Aspergillus niger* spores equals 101°, and that of polyethylene and cellophane equals 93.6° and 20.6°, respectively.

According to the famous Debroigne rule [267], adhesion forces are maximal at the same polarity of the surfaces, i.e. at adhesion of hydrophobic surface to hydrophilic one and vice versa.

In case of meeting this rule for the studied objects, one may expect more intense adhesive interaction between *Aspergillus niger* spores and polyethylene rather than cellophane. Carried out investigations (Table 4.5) show that in the present case, the adhesion force of *Aspergillus niger* to cellophane is almost three-fold and index K 1.8-fold above the appropriate indices for polyethylene. Analogous results have also been obtained for other studied species of microscopic fungi. It has been found that the hydrophilic-hydrophobic balance of their spores is shifted towards hydrophoby (θ falls within the range of 120° - 90°). Therewith, for the adhesive interaction with cellophane, the average adhesion force, F_{∞}^{50}, and the index, K, significantly exceed the appropriate values for polyethylene.

Oppositions of the data obtained to the Debroigne rule can be explained based on the structure and chemical composition of the external layer of the fungus spore surface, its inhomogeneity (patchiness) and ability to vary during adhesion.

It has been shown [129, 256] that concentration of phospholipids determining cell hydrophoby at its contact with a substrate is abruptly reduced from ~70% to 30%, and a protein is intensively produced by the microorganism, which shifts the hydrophobic-hydrophilic balance of the cellular system towards hydrophoby. Such mechanism stipulates variation of the hydrophilic-hydrophobic balance of the cell surface due to the appropriate balance of the substrate that provides its strong fixation to various (by hydrophoby) surfaces. One may suggest that the degree and intensity of the property variations of the spore surface will be determined by both biological features of the current microorganism and the substrate (material) properties.

It has been suggested [268] that water content in polymers is one of the key factors, on which the quantity of biomass formed on the material surface during growth of microscopic fungi depends. Obviously, the material parameter mentioned should also be significant for processes of adhesion.

Table 4.5 shows values of water sorption (Q) of the studied materials, determined experimentally in accordance with GOST 12020-72 [269] at temperature 20°C and 24-hour exposure of the polymer samples in water.

Analysis of these data allowed ascertaining correlation between Q and indices of adhesive interaction, F_{∞}^{50} and K. The rate of adhesive bond formation and, consequently, the average maximal adhesion force are reduced with water sorption. This connection is illustrated by graphs shown in Figure 4.8. Clearly in semi-logarithmic coordinates, ln$Q - F_{\infty}^{50}$ and ln$Q - K$, points corresponded to indices of the adhesive interaction between *Aspergillus niger* spores and different materials can be approximated by a linear analytical model testifying about the existence of direct proportionality between the adhesion parameters and ln$Q$.

Hence, the results obtained allow a supposition that adhesion of microscopic fungus spores to polymeric materials depends on water content in them. The present regularity can be used for preliminary estimation (forecasting) of the appropriate adhesive interactions.

Table 4.6 shows adhesion indices for spores of various species of microscopic fungi to polyethylene and cellophane. Studied species of fungi are places in the Table in the descending order by the average adhesion force of their spores to materials. Hence, values of the rate constant of adhesion force formation increase in this sequence of microscopic fungi.

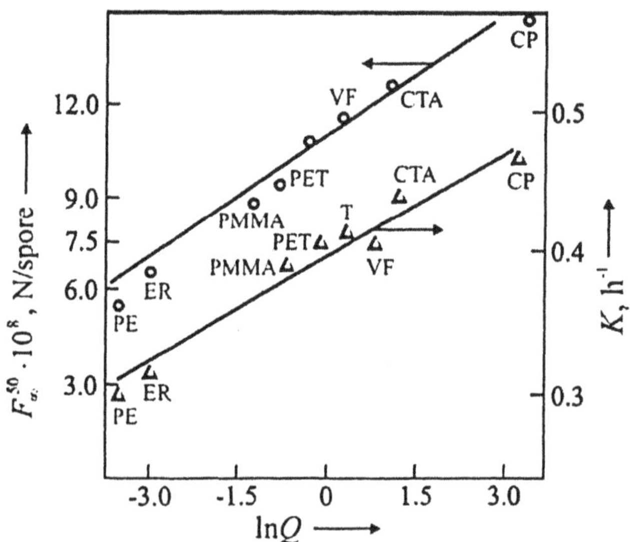

Figure 4.8. Dependencies of the average adhesion force, F_{∞}^{50}, and the rate constant, K, of adhesive interaction formation on water sorption of polymeric materials (notations of the materials are present in accordance with Table 4.5)

Moreover, analysis of the data from Table 4.6 shows that the series obtained of adhesive ability of spores of various fungus species to polyethylene and cellophane coincide. This may testify about existence of a property (parameter) general for spores of all studied fungal species, which cause the determining influence on their adhesive interaction both with hydrophobic and hydrophilic surfaces.

It is common knowledge [112] that, all other factors being the same, adhesion of dust and powders of non-biological origin is frequently defined by square of adhesive contact, which, in its turn, generally depends on size (radius) of the particles. Table 4.6 shows the values of average radii of microscopic fungus spores, r, and Figure 4.9 shows dependencies of the adhesion indices on their values. Clearly, there is an interconnection between the values of r and indices F_{∞}^{50} and K reflecting comparative ability to adhere various species of fungi. The average force of the spore adhesion to polyethylene and cellophane is reduced with the radius, and K increases simultaneously. It is noted [132] that the smaller are microbiological objects, the more intensive and faster changes in chemical composition of their cell walls end. One may suggest that such changes responsible for adhesion of the spores obey the mentioned regularity. This explains the obtained experimental dependencies $K = f(r)$. The smaller are geometrical sizes of the spores (their radius r), the higher are the rate of formation of adhesive forces and the index K characterizing it.

Table 4.6

Indices of the adhesive interaction of microscopic fungus spores with polyethylene and cellophane. Temperature - 29°C, humidity – 98%

Species of microscopic fungi	Mean radius of spores, μm	Polyethylene		Cellophane	
		$F_{\infty}^{50}\cdot 10^8$, N/spore	K, hour^{-1}	$F_{\infty}^{50}\cdot 10^8$, N/spore	K, hour^{-1}
Aspergillus niger	4.2	5.1	0.27	14.2	0.46
Paecilomyces varioti	3.6	4.8	0.30	12.8	0.49
Penicillium chrysogenum	2.6	3.0	0.32	7.7	0.51
Aspergillus flavus	2.3	2.5	0.32	6.2	0.53
Penicillium cyclopium	1.6	0.5	0.54	1.6	0.94
Aspergillus terreus	1.2	0.38	1.05	0.7	1.50

As mentioned above (Section 4.1), variation of the adhesion force with time is associated with increase of the contact square between spores and the surface. In this case, basing on conditionally accepted geometrical shape of spores, the contact square should be proportional to $r^{2/3}$. This means that the spores with greater radii, as a consequence, will give greater squares of the adhesive contact and greater force of adhesion to the surface. This is the probable reason of experimentally determined symbate increase of r and F_{∞}^{50} values.

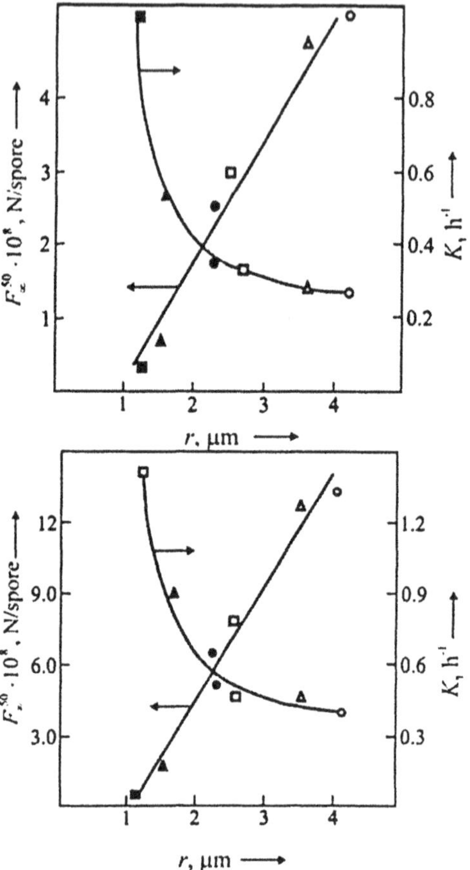

Figure 4.9. Dependence of the average adhesion force (F_{∞}^{50}) and the rate constant (K) on the average radius of fungal spores (r).
a – polyethylene; b – cellophane;
○ – *Aspergillus niger*; Δ – *Paecilomyces varioti*; □ – *Penicillium chrysogenum*; • – *Aspergillus flavus*; ▲ – *Penicillium cyclopium*; ■ – *Aspergillus terreus*

Analysis of the data shown in Table 4.6 indicates also that for *Aspergillus terreus* (r = 1.2 μm), differences in appropriate values of the adhesive indices for adhesion to polyethylene and cellophane are 70% and 42%, respectively. For spores of *Aspergillus niger* (r = 4.2 μm), these differences are much higher and give 180% for F_{∞}^{50} and 70% for the rate constant K.

One may suggest that the feature mentioned is also stipulated by sizes (radii) of the spores of microscopic fungi. For small conidia, the contact square is also small, due to which the surface origin should less tell upon the values of F_{∞}^{50} and K parameters. The contact square increases with the spore size. As a consequence, the influence of polymeric surface becomes more and more valuable that causes rise of differences in the appropriate indices for polyethylene and cellophane.

It is common knowledge [112] that the ratio of sizes of finely dispersed particles and irregularities (rough edges) of the surface may also affect the square of their contact and adhesive interaction, respectively.

The authors of the present monograph have studied adhesion of *Aspergillus niger* spores to surfaces with different degree of roughness. PMMA samples were used, treated by flint-paper of various abrasive grain. The sample surface roughness was determined on the profiler Talysurf-10.

In accordance with GOST 2789-73 [270], roughness was estimated by irregularities of the cross-section profile of the real surface by a plain and characterized by the following parameters: mean arithmetic deviation of the profile (R_a), height (R_z) and average step (S) of the profile irregularities by apexes.

Table 4.7 shows values of the F_{∞}^{50} index for adhesion of *Aspergillus niger* spores to PMMA surfaces with different degrees of roughness.

Table 4.7

Investigation results of the PMMA roughness effect on adhesion of *Aspergillus niger* spores (r = 4.2 μm). Temperature – 29°C, humidity – 98%.

Surface profilograms	Surface roughness, μm			$F_{\infty}^{50} \cdot 10^8$, N/spore	Schemes of contact between spores and surface
	R_a	R_z	S		
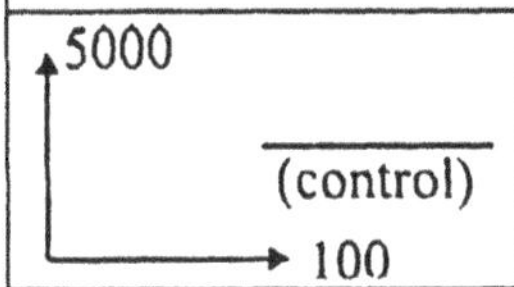 5000 (control) 100	0	0	0	8.5	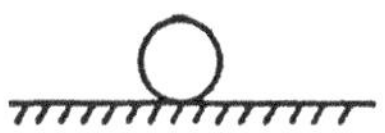
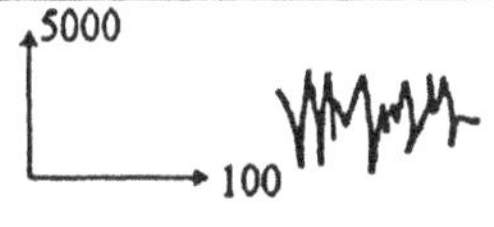 5000 100	1.1	15	4.0	6.2	
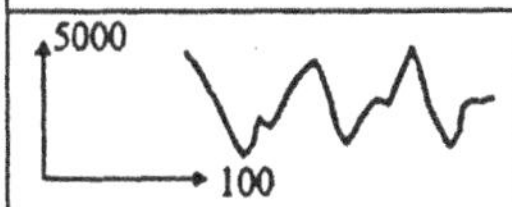 5000 100	4.0	40	20	9.8	

The obtained data allow separation of three possible cases of support roughness effect on the adhesion of fungal spores. The first case characterizes interaction between the spores and the smooth (abrasive untreated) surface of PMMA (the control sample). The second case is realized when the average step of the profile roughness by apexes (S) is smaller than the radius of spores. Hence, the data from Table 4.7 show that the real contact square between adhering objects increase. This probably causes the observed decrease of the F_{∞}^{50} value comparing with the one obtained in use of samples treated by an abrasive (control sample).

In the third case, the surface roughness increases the adhesion force of *Aspergillus niger* spores. Hence, the average step of roughness is comparable with the size (radius) of the spores which, apparently, provides for the contact square increase of adhering objects.

Hence, the surface roughness affects the adhesion contact square of both particles of non-biological origin and spores, being able to increase or reduce the force of their adhesion to the surface.

Detected dependencies of adhesion on properties of the surface and fungal spores indicate that the determining factors of this process are water capacity and roughness of the material, and the average radius of the spores. Regularities obtained allow comparative estimation and forecasting of studied adhesive interaction under some environmental conditions.

To estimate possibility of parameters F_{∞}^{50} and K, application to forecasting of real adhesion, regularities of changes in the mentioned indices in the case of test temperature and humidity mode variation have been determined. It is common knowledge that under natural conditions, temperature and humidity of the environment cause the determining effect on occurrence and intensity of biodamaging of materials (articles).

Figure 4.10 shows dependencies of adhesive interaction indices of *Aspergillus niger* spores to polyethylene and cellophane on temperature (T) and air humidity (φ). Clearly F_{∞}^{50} and K indices are sensitive to variable factors of the environment. Hence, regularities of the temperature-humidity complex effect on adhesion of spores are general for both hydrophilic and hydrophobic materials. Maximal adhesion force and the rate constant of its formation are observed at 29°C and 98% air humidity, i.e. under the most favorable conditions for the growth of microscopic fungi [132, 183]. As temperature increases from 15 to 29°C (at constant air humidity), K index is changed by 5 – 7% only, which is within the error of its determination. The effect of temperature on the average force of spore adhesion to polyethylene and cellophane is also insignificant. At the same time, $F_{\infty}^{50} = f(T)$ dependences shown in Figure 4.10 testify about a tendency of F_{∞}^{50} decrease with temperature raise from 15 to 29°C. Hence, temperature conditions of the adhesion contact (at constant humidity) are insignificant for the state of F_{∞}^{50} and K indices.

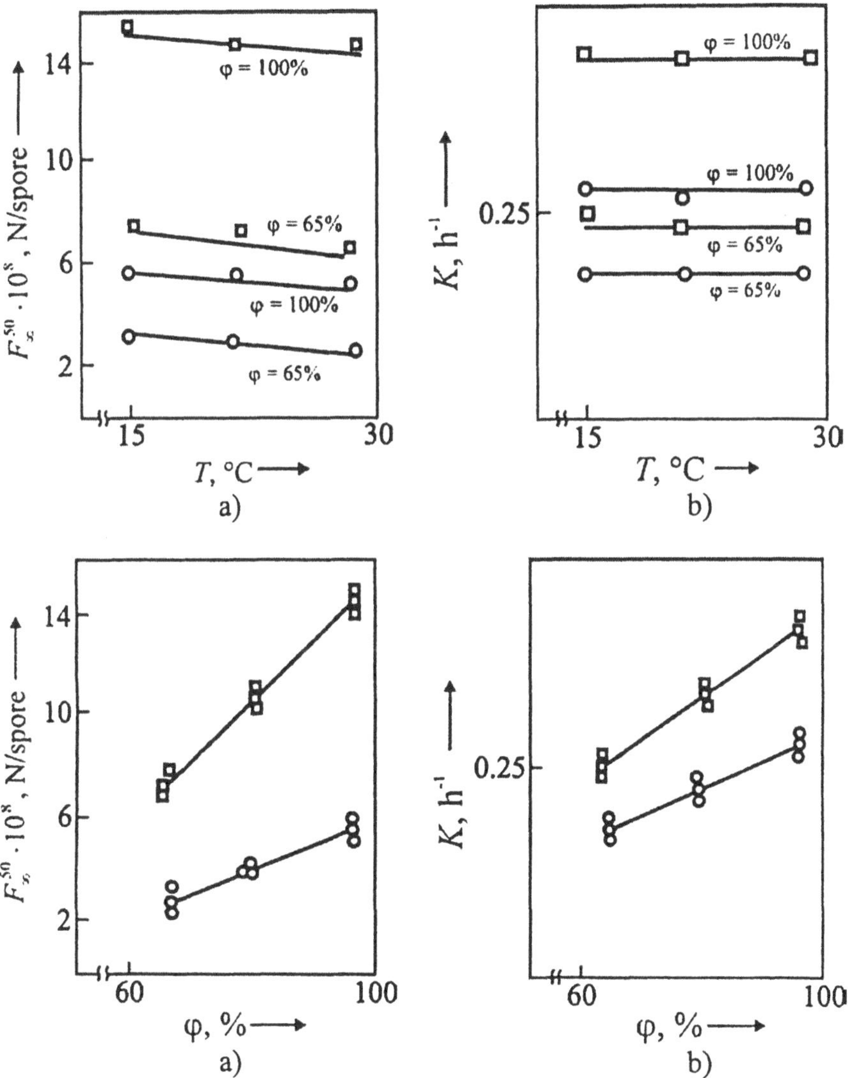

Figure 4.10. Dependencies of indices F_{∞}^{50} (a, c) and K (b, d) on temperature and humidity of the adhesive interaction between *Aspergillus niger* spores and polyethylene (○) and cellophane (□)

On the contrary, the data shown in Figure 4.10 testify that adhesion indices depend on the air humidity. Humidity raise is accompanied by a significant increase of F_{∞}^{50} and K values. Calculations have indicated that the experimental dependencies $F_{\infty}^{50} = f(\varphi)$ and $K = f(\varphi)$ can be approximated by linear equations shown in Table 4.8.

For *Aspergillus niger*, adhesion to polyethylene and cellophane, values of humidity coefficients α_F and α_K are $10.0 \cdot 10^{-8}$ N/spore (α_F), $0.28r^{-1}$ (α_K) and $58.0 \cdot 10^{-8}$ N/spore, $0.52r^{-1}$, respectively.

The analysis of obtained values shows that these indices are higher for cellophane rather than for polyethylene. E.g. adhesion of spores to hydrophilic surface is more sensitive to the humidity conditions of the environment. Hence, for both hydrophilic and hydrophobic materials, variations of air humidity cause more significant influence on adhesion forces than on the rate of their formation.

Table 4.8

Analytical dependencies of adhesion indices of *Aspergillus niger* spores to polyethylene and cellophane on air humidity

Adhesive interaction index	Type of equation
F_{∞}^{50}	$(F_{\infty}^{50})_{\varphi} = (F_{\infty}^{50})_{\varphi 1} \pm \alpha_F(1 - \varphi)$, (4.5)
K	$K_{\varphi} = K_{\varphi 1} \pm \alpha_K(1 - \varphi)$, (4.6)

Note: $(F_{\infty}^{50})_{\varphi}$ and K_{φ} are values of F_{∞}^{50} and K indices at air humidity φ ($0.65 \leq \varphi \leq 0.98$); $(F_{\infty}^{50})_{\varphi 1}$ and $K_{\varphi 1}$ are values of F_{∞}^{50} and K indices air humidity φ_1, for example, $\varphi_1 = 98\%$; α is the humidity coefficient. As $\varphi < \varphi_1$, the right part of equations (4.5) and (4.6) represents a sum, and at $\varphi > \varphi_1$ – a difference of appropriate values.

Thus one may conclude that indices F_{∞}^{50} and K and correlation equations (4.5) and (4.6) allow characterization and forecasting of adhesive interaction between the fungal spores and the surfaces at any combination of temperature and air humidity in the studied range of measurements. The determining influence on adhesion of microscopic fungus spores to polymeric materials is induced by humidity of the environment.

Generally, determined dependencies of the adhesion indices, F_{∞}^{50} and K, on water capacity and roughness of the material, radius of fungal spores and the temperature-humidity complex of factors allow comparison and forecasting of the adhesive interaction of particular microorganism-material couples at various temperatures and humidity values of the environment.

Deduced analytical dependencies (4.4) – (4.6) are of the empirical type that restricts capabilities of their application by investigated materials and microscopic fungi only. At the same time, the objects of the current investigations were various species of fungi, as well as polymeric materials differed by chemical composition, structure, and components (fillers, dyes, etc.). Therewith, community of macro regularities of the adhesion formation was determined for all studied adhesive couples. This allows a supposition that the forces of adhesion of fungus spores to polymeric surfaces are of the same origin. Then the suggested approach to investigation and estimation of the microbiological adhesion, as well as detected regularities can be applied to much wider spectrum of materials and species of fungi.

The data presented by the authors of the current monograph allow some suggestions about the origin of forces determining formation of the real adhesive interaction between microscopic fungus spores and polymeric materials. It is common knowledge [112, 252] that adhesion of non-biological particles as well as bacterial cells can be caused by various forces. The most valuable among them are molecular, chemical, electric and capillary forces.

The reaction rate and yield increase with temperature of the reaction system is typical of chemical processes, including many biochemical ones. It has been found by the authors (Figure 4.10) that temperature variations cause an insignificant effect on the indices, F_{∞}^{50} and K. Hence, a tendency to increase of the average adhesion force with temperature decreasing from 29 to 15°C has been observed. Such character of the temperature dependence is peculiar to physical processes of molecular, electric and capillary interactions.

Under high humidity conditions (over 65%), the presence of the phase water film between contacting surfaces of spores and polymer is quite probable. In this case, any significant effect of electrical forces occurring at direct contact between the bodies due to proceeding of donor-acceptor processes on the adhesive bond is practically excluded. Moreover, the presence of water promotes electric conductivity increase of the contact zone that should seriously decrease Coulomb's forces and reduce time of their interaction. Hence, one may suggest that for test modes used, electric forces do not cause an impact on formation of adhesive interaction between the fungal spores and the surfaces. Detected high adhesive capability of the spores in relation to both hydrophilic and hydrophobic surfaces has allowed a hypothesis about participation of microbial cell metabolites at their contact with substrate in formation of the adhesive bonds. In the frames of this hypothesis, above all, adhesion is provided by molecular forces. At the same time, besides molecular forces, capillary ones cause a significant (in some cases, dominating) effect on the adhesion formation at high air humidity. These forces occur due to condensation of liquid in a gap between interacting surfaces.

Ref. [112] gives theoretical proof and experimental confirmation that molecular and capillary forces of adhesion vary symbately with the change of radius of finely dispersed particles adhered to the surface. The authors of the monograph have found that the spore adhesion force to polymeric surface increases monotonously with their radius (Figure 4.9). The presence of such dependence confirms the determining effect of molecular and capillary forces on formation of the real adhesive interaction studied.

Capillary forces is so higher, the higher is wettability of interacting surfaces by liquid condensed on them. Obviously, high adhesion forces to hydrophilic surfaces can be associated with higher capillary forces occurring at the interface of three condensed phases – polymer, fungus spore and liquid (water or aqueous solution of the substance of the external cell wall responsible for adhesion).

Hence, the analysis carried out and the experimental data obtained give bases to a supposition that under temperature-humidity conditions studied adhesion of microscopic fungus spores to polymeric materials is stipulated by physical processes of molecular interactions between the surfaces and capillary condensation of liquid in the gap between them.

Experimental results obtained (Tables 4.5 – 4.8) illustrate a possibility of using indices F_{∞}^{50} and K for determination of the efficiency of some methods reducing adhesion forces: reduction of water absorbing capacity of the material, increase of its water repellency, variation of the surface roughness and external temperature-humidity conditions of the contact.

The impact of biocides introduced to the material composition was also studied. The object of these investigations were special samples of varnished fabric prepared by applying nitrocellulose varnish ETs-959 with 5% biocinoma /?/ biocide injected into it on a fabric support.

It has been found that introduction of the mentioned biocide causes no effect on the adhesive ability of the varnished fabric to *Aspergillus niger* spores. Obviously, biocidal action of this substance is displayed at its longer contact with the microorganism. One may suggest that the influence of compounds possessing biocidal properties on adhesion is determined not by biochemical features of cell metabolism inhibition, but above all, by the feature of these substances to change physicochemical properties of the surface, which are responsible for formation of the adhesive bonds.

Hence, the experimental data obtained allow a conclusion that the indices F_{∞}^{50} and K can be used for development and comparative estimation of methods regulating adhesion of microscopic fungus spores to surfaces. The possibility of applying these indices to estimation of the biocide influence on adhesion requires additional investigations.

CONCLUSION

It has been found that common regularities typical of the adhesive interaction between finely dispersed particles of non-biological origin and solid surfaces, are peculiar to the adhesion of spores of microscopic fungi to polymers. Adhesion force intensities of spores of even a single fungus species to the surface are different and depend on the contact square of the adhering objects.

It has been shown that for sedimentation of spores from calm air to a material, the adhesion force exceeds $1.8 \cdot 10^{-12}$ N/spore independently of the material origin and species of microorganisms. In this case, all supplied spores are adhered to the surface, and kinetics of variation of their number on the surface submit to the equation of finely dispersed particles' precipitation (the Stokes law) and depends on the radius of spores and sample tilt to the spore flux direction.

Distribution of the fungal spore adhesion force to the surface submits to the normal law. The factor determining this type of adhesion force distribution is dispersion of the spore radii, which also submit to the normal distribution law.

The adhesion force increases with time of spores' contact with the surface. Kinetics of the adhesion force variation submits to the exponential equation – the adhesion equation. Parameters of this equation most fully characterizing adhesive interaction between microscopic fungus spores of and surfaces are as follows: mathematical expectation of the adhesion force (mean value of the adhesion force) after the end of adhesion force formation of all spores present on the surface (F_{∞}^{50}) and the rate constant of adhesion force formation of spores to the surface (K). These parameters indicate the initial stage of microbiological damaging of materials.

The determining technique for adhesion indices has been proved experimentally. They are calculated on the basis of dependencies of the adhesion number on detachment force obtained in experiment at various durations of the microbial cell contact with materials.

Approbation of adhesion indices and the technique during investigation has displayed their appropriateness for estimating and forecasting capabilities of materials and microorganisms-destructors for the adhesive interaction and efficiency of control measures for this interaction (protection means).

Quantitative interconnections between normal water absorption of polymeric materials, roughness of the surfaces, radii of spores of biodestructors and indices F_{∞}^{50} and K have been detected. Water absorption increases with the adhesion force of spores and the rate of its formation. Increase of the spore radius induces rise of the adhesion force and simultaneous reduction of the rate constant of the process. Surface roughness increase may reduce or increase the adhesion force of spores to the surface.

It has been shown that under favorable conditions for microscopic fungus growth ($T > 15°C$ and $\varphi > 65\%$), the force and the rate constant of adhesion are mainly determined by humidity (φ) and are insensitive to the temperature mode. Analytical dependencies binding adhesion indices (F_{∞}^{50} and K) and φ value have been deduced.

Chapter 5.

Growth of microorganisms-destructors on materials

Adhesive interaction between microbial cells and surfaces stipulates primary contact between these objects creating necessary initial conditions for microbiological damaging of material. In accordance with the assumed stage-by-stage approach to investigations, the next stage of the process is further development of microorganisms which is their growth on the material.

The present Chapter shows investigation results of kinetics of biodestructor growth on various materials. Analytical models, indices of this stage of biodamaging, techniques for their determination, as well as the possibility of their use for estimating and forecasting microbial growth, efficiency of protection measures and methods are proved.

5.1. QUANTITATIVE DESCRIPTION AND INDICES OF THE PROCESS

Kinetic curves of dry specific biomass growth (Chapter 2) of microscopic fungi and bacteria on polymers and in combustible materials (fuels) and lubricants are shown in Figure 5.1.

Clearly every microorganism-material couple at any particular moment of the biomass growth is characterized by the self value of the latter (m_t). Maximal volume of the biomass obtained in experiment and disposition of growth kinetic curves in (m; t) coordinates are also individual parameters. These curves are characterized by different time intervals since microbial cells are applied to the material till initiation of the biomass change and the curve tangents to the axis of time, i.e. growth rates of the biomass. Consequently, the parameter used which is the specific dry biomass, and kinetic curve of its change allow consideration of both degree of the biomass growth (value of the biomass at any particular moment) and the growth rate of the biomass on materials.

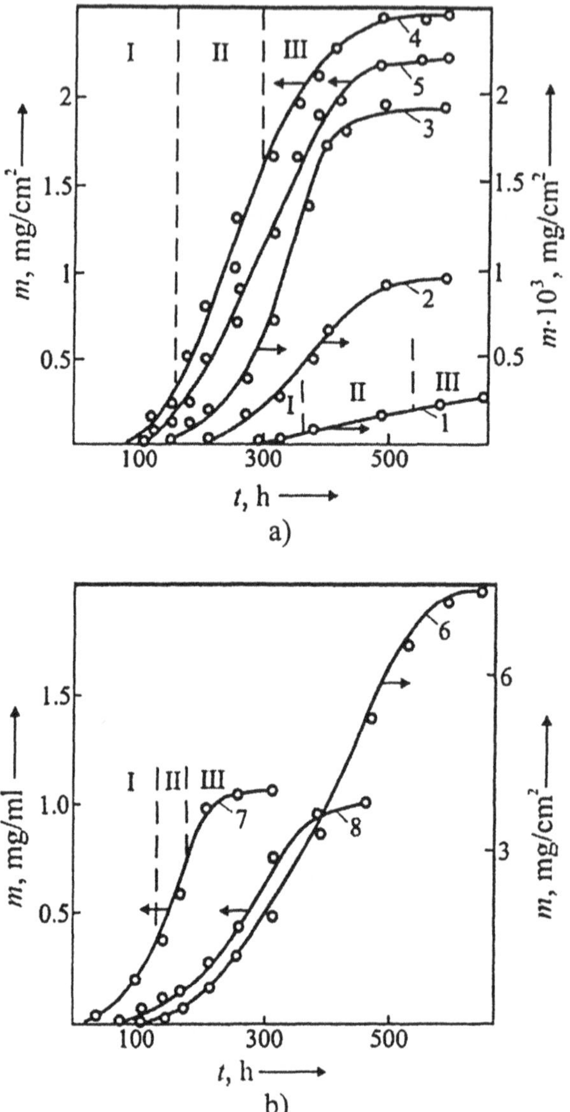

Figure 5.1. Dependence of the biomass quantity (m) of microorganisms on time of their cultivation on polymeric materials (a) and fuels and lubricants (b)

Aspergillus niger biomass growth on PE (curve 1), PVC (2), PMMA (3), varnished fabric (4); biomass growth of *Penicillium chrysogenum* on varnished fabric (5) and *Cladosporium resinae* in TS-1 (6), *Pseudomonas aeruginosa* in TS-1 (7), and *Bacillus sp.* in MN-7,5U (8)

All experimental kinetic curves $m_t = f(t)$ are similarly S-shapes, which is typical of the development of various biological objects [132, 161 – 163, 271, 272]. In particular, the S-shape is peculiar to the graphs of kinetic curves of the biomass growth and increase of the quantity of bacterial and microscopic fungus cells in the limited vital space, the so-called "periodical (regular) culture" [132, 133]. Growth of microorganisms in such "closed system", where the quantity of nutrition is limited and elimination of metabolites is difficult, displays many signs similar to the process studied.

One may suggest that the regularities typical of such systems will also be true for growth of microorganisms on technical materials, usually present under similar conditions.

As mentioned in Chapter 1, several basic phases (stages) of microorganisms' development in a periodical culture are differed [132, 133]. Differing by the type, obtained parts of the kinetic curves of biomass growth (Figure 5.1) correlate well with these stages. The first part (part 1, Figure 5.1) corresponds to the so-called adaptive stage (of the lag-phase), which includes the period of accelerated irregular growth of the biomass. Duration of this phase is defined by the time interval since inoculation of the nutritious substrate (material) till reaching the maximal acceleration of the growth rate. During some time (part 2, Figure 5.1), biomass is monotonously increased at a maximal constant rate – the exponential phase. After that the intensity of metabolic processes is reduces gradually, which is accompanied by reduction and then termination of the biomass growth – the stationary phase (part 3, Figure 5.1).

Mathematical models of growth of biological objects are usually (Chapter 1) based on solving differential equation as follows:

$$\varphi(\tau) = \mathrm{d}F/\mathrm{d}\tau(F),$$

where $\varphi(\tau)$ is the specific rate of biological system growth; F is a growth parameter; τ is time.

Table 5.1 shows dependencies $\varphi(\tau)$ the most frequently used for description of the growth kinetic curves and analytical solutions of the mentioned differential equation for these functions.

Analytical model most properly reflecting kinetics of the biomass growth on materials was selected by a regressive analysis. Values of m_t, obtained in kinetic experiment, were used as parameters of culture growth on the material. The growth time (τ) was assumed equal to the difference between cultivation time (t) corresponded to of m_t value and time (L) since the moment of sample poisoning till the initial change of biomass, detected during the experiment.

Table 5.1

Analytical solutions of differential equation on growth in biological systems

Type of $\varphi(\tau)$ dependence	Analytical solution of the growth equation	Correlation index
$\varphi(\tau) = a\exp(-b\tau)$	$F_\tau = F_\infty \exp\left[-\frac{a}{b}\exp(-b\tau)\right]$ (5.1)	$0.30 < R_{m,\tau} < 0.52$
$\varphi(\tau) = \dfrac{b\exp(-b\tau)}{1/a - \exp(-b\tau)}$	$F_\tau = F_\infty[1 - a\exp(-b\tau)]$ (5.2)	$0.35 < R_{m,\tau} < 0.78$
$\varphi(\tau) = \dfrac{b}{1 + \dfrac{\exp(b\tau)}{a}}$	$F_\tau = \dfrac{F_\infty}{1 + a\exp(-b\tau)}$ (5.3)	$0.85 < R_{m,\tau} < 0.99$

Note: a and b are parameters; F_τ and F_∞ are values of biological object growth parameter and its border value (at $\tau \to \infty$), respectively.

According to the experimental data, values of parameters a and b were selected by the least squares method for each of equations (5.1) – (5.3). The signs of acceptability of the equation describing the growth process studied were standard error and correlation index values.

Calculations have shown (Table 5.1) that all experimental dependencies $m_t = f(t)$ are described by equation (5.3). Application of this equation gives minimal standard errors and maximal correlation indices, $R_{m,t}$. Hence, determination coefficients $B_{m,t} = R^2_{m,t}$ fall within the range of 0.76 – 0.92 for various microorganism-material couples.

Thus microscopic fungus and bacteria biomass increase during their cultivation on the materials (for $t \geq L$) obeys the kinetic equation as follows:

$$m_t = \frac{m_\infty}{1 + a \cdot \exp[-b(t - L)]}, \qquad (5.4)$$

where m_t is the specific dry biomass at time t of microorganism cultivation on the material; m_∞ is the border (maximal) value of m_t obtained in the experiment; L is the time period, during which the biomass of microbial cells poisoning the material remains unchanged; a and b are parameters of the biomass growth.

Equation (5.4) reflects quantitative interconnection and considers features of proceeding of all phases forming the circle of the microorganism growth on materials. That is why it seemed desirable to consider the physical sense of parameters from this equation in more detail to select quantitative indices of the biodamages studied.

As mentioned above, parameter L characterizes duration of the phase of microorganism adaptation to the material (the lag-phase). The smaller is L, the shorter is the lag-phase and the more active is proceeding of the microorganism-material interaction preceding biomass increase and, obviously, the higher is the material ability to support the current microorganism by nutritious elements and energy, necessary for its growth.

It has been already mentioned that the lag-phase (L) also includes the period of accelerated irregular growth of biomass (L_1). In this case, total duration of the lag-phase (L_2) is determined as follows:

$$L_2 = L + L_1. \tag{5.5}$$

As the period of accelerated irregular growth is terminated, which also means termination of the lag-phase, acceleration of the biomass growth reaches its maximum typical of the next phase of microorganism development which is the exponential phase.

Analysis of equation (5.4) shows that graphically the moment of transition from the lag-phase to the exponential phase corresponds to the point of the first extreme on the kinetic curve of acceleration of the biomass growth: $\dfrac{d^2m}{d(t-L)^2} = f(t-L)$. The gradient of acceleration is maximal in this point. Then duration of accelerated irregular growth can be determined from the condition as follows:

$$\frac{d^3m}{d(t-L)^3} = 0.$$

Differentiating equation (5.4) and carrying out appropriate transformations, we obtain:

$$L_1 = \frac{\ln(a) - 1.31}{b}. \tag{5.6}$$

Using experimentally determined value of L and calculated values of L_1, one can easily determine "real" duration of the lag-phase (L_2) by equation (5.5).

Such method of the lag-phase calculation should allow reduction of mistake of its determination associated with the instrumental error of the biomass registration method and arbitrary selection of periodicity of its measurements during the experiment. In this connection, besides parameters of

L, in the case of comparative estimations of microorganisms' growth on materials, for example, as well as for the data obtained by various experimental methods of biomass determination, it is desirable to use values L_1 and L_2 calculated by formulae (5.5) and (5.6).

Maximal biomass m_∞, formed on the material during growth of microorganisms, and parameter a are linked with one another by the expression, obtained from equation (5.4) under the condition $t = L$:

$$a = \frac{(m_\infty - m_0)}{m_0}, \tag{5.7}$$

where m_0 is the initial value of biomass growing on the material.

Expression (5.7) shows that the parameter a (as well as m_∞) characterizes some result of the growth process, which is finite under experimental conditions. If m_∞ represents the value of the maximal biomass, then a parameter characterizes the degree of its possible increase in relation to the initial quantity of biomass, m_0, capable of growing on the material. The greater are m_∞ and a parameters, the higher, obviously, is the ability of the interacting microorganism-material couple to provide for biomass growth.

According to equation (5.4), the expression $a \cdot \exp[b(t - L)]$ determines the growth rate. Hence, the higher is the pre-exponential coefficient b, the faster (at the current value of parameter a) the growth will proceed and the maximum of biomass (m_∞) will be reached. Thus, parameter b represents the rate constant and characterizes the specific growth rate (the growth rate of the specific biomass).

Table 5.2 shows values of the growth equation (5.4) parameters for some microorganism-material couples studied. Clearly the self selection of parameters is peculiar to every such pair. Their values change both at varying the material type (at growth of particular species of microorganism on it) and microorganism species (at its propagation on current type of the material). As a consequence, parameters of equation (5.4) are sensitive both to the material origin and species features of the microorganisms developing on them.

A stable connection between parameters b, m_∞ and L (L_2) exists for various couples of interacting objects. Parameters b and m_∞ are changed symbately. They increase with shortening of the lag-phase duration (or L_2 value). Such connection correlates with the above-considered suggestions about physical meaning of these parameters, based on the notions about materials as nutritious medium for microorganisms. The higher is material capabilities for supplying biological growth, the shorter the phase of microorganism adaptation

(L) to it should be and the greater is the maximal biomass (m_∞) and the growth rate (b).

Table 5.2

Values of parameters in equation (5.4) for the growth of microorganisms on material

Material	Microorganism	m_∞, mg/cm^2 (mg/ml)	$b \cdot 10^2$, hour^{-1}	a	m_0, mg/cm^2 (mg/ml)	L, hour	L_2, hour
Polyethylene	*Aspergillus niger*	0.00027	1.5	12.5	0.00002	288	369
	Penicillium chrysogenum	0.00018	1.1	1.7	0.00001	312	450
Varnished fabric	*Aspergillus niger*	2.5	2.2	34.7	0.07	96	198
	Penicillium chrysogenum	2.2	1.9	30.4	0.07	120	227
Diesel oil TS-1	*Cladosporium resinae*	7.5	2.9	61.1	0.12	96	192
	Pseudomonas aerugenosa	1.08	3.8	11.0	0.09	24	56
Oil MN-7,5U	*Bacillus sp.*	1.0	3.4	9.0	0.10	72	95

The data obtained (Table 5.2) show also that the quantity of initial growing biomass (m_0), calculated by equation (5.7), is different for different microorganism-material couples. At the same time, samples of various materials were polluted by the same quantity of viable spores during the tests (Chapter 2) and, consequently, the quantity of initial (primary) biomass was approximately the same on such samples. One may suggest that not all microbial cells applied to samples of materials are capable of forming biomass.

Of special attention is the fact that the value of m_0 is so higher, the higher are parameters m_∞ and b and the lower is L (L_2). Such type of changes in these parameters is also contradictory to their physical meaning. The more nutritious medium the material is for microorganisms (i.e. the higher are parameters m_∞ and b and the lower is L), the greater part of the biomass applied to it should be capable of growth.

The self value of parameter a, according to the above-mentioned characterizing the relative biomass increase, corresponds to every studied material. For the majority of the microorganism-material couples shown in Table 5.2, the following relation between this parameter and other indices of equation (5.4) is observed: values of parameters m_∞, m_0 and b are increased and the lag-phase duration is decreased with the parameter a increase. At the same time, growth of microscopic fungi on polyethylene displays somewhat different tendency. As *Aspergillus niger* species is cultivated on this polymer, parameter a of the process is somewhat lower than for *Penicillium chrysogenum* growth.

On the contrary, analysis of other parameters of equation (5.4) measured for these two couples show more intense growth of *Aspergillus niger* biomass rather than *Penicillium chrysogenum*. Obviously parameter a is more complexly, ambiguously associated with other parameters of equation (5.4) and growth as the entire process.

Hence, investigations performed have proved theoretically and experimentally the physical meaning of kinetic equation parameters describing changes of microorganisms' biomass during their growth on technical materials. The parameters characterize the degree and the rate of the current growth and consider its features for various microorganism-material couples. This stipulates the possibility of using parameters of equation (5.4) as quantitative indices of the studied stage of microbiological damage to materials.

The regular type of changes in the parameters for different microorganism-material couples has been detected. The indices m_∞, m_0 and b change symbately. Their increase is accompanied by shortening of the lag-phase duration, L. This allows the use of growth indices not in a complex but separately, for example, for solving some particular tasks or performing various estimations. In some cases, such approach will allow a significant reduction of experimental studies.

Determination of the growth indices supposes experimental obtaining of the dependence of growth quantitative parameter (dry specific biomass) on time of the microorganism cultivation on the material and calculation of parameters b, L (L_2), m_∞, m_0 and a based on these experimental data and equation (5.4).

5.2. DEPENDENCE OF MICROBIOLOGICAL GROWTH ON PROPERTIES OF MATERIAL, MICROORGANISM AND TEMPERATURE-HUMIDITY CONDITIONS OF THE ENVIRONMENT

The possibility of applying suggested indices to comparative estimations of the abilities of materials and microorganisms to provide for growth of the latter and the methods of its control and to development of forecasting methods has been tested. The same methodological approach has been applied as for carrying out appropriate studies of the adhesive interaction between microbial cells and materials. The functional suitability of the parameters was judged by adequacy of the results obtained on their basis to known common connections between the microbiological growth and parameters of objects participating in it (material, microorganism) and the environment.

It has been mentioned in Chapter 1 that possibility and intensity of growth are determined by properties of the material as nutritious medium, biochemical properties of microorganism and in many respects by the temperature-humidity conditions of cultivation.

Quality of materials as nutritious media is mostly associated with concentrations of water and components suitable as organic (carbonic) alimentation for microorganisms. That is why the materials have been characterized by normal water absorption (Q) and relative concentration of components, assimilated by microorganisms during the analysis.

The value of Q has been determined in accordance with ref. [269]. Concentration of organic nutritious substances in the material has been varied using special samples of varnished fabric. When prepared for tests, the samples have been thermally oxidized during different times. It is commonly known [273, 274] that the treatment mentioned causes degradation of varnished fabric, accompanied by accumulation of low-molecular fragments of macromolecules in it. Such fragments are easily assimilated by microorganisms [2, 275]. This allowed the use of preliminary thermooxidation duration as a parameter reflecting indirectly the concentration of components assimilated by microorganisms in the varnished fabric. The longer the period of thermooxidatrion is, the higher concentration of such components in the material should be.

Table 5.3 shows parameters of *Aspergillus niger* microscopic fungus growth on polymeric materials and standard water absorption of the latter. Clearly the majority of studied *Aspergillus niger*-material couples display proceeding of the regularity, already mentioned in the previous Section. Parameters m_∞, b, a, and m_0 vary symbately. Their increase is accompanied by shortening of the lag-phase duration.

Simultaneously, it should be noted that thermooxidation of varnished fabric (carried out to change the quantity of carbonic alimentation in it) does not induce any significant variation of the specific rate of biomass growth, m_0 value and duration of microorganism adaptation to the material (the lag-phase). Obviously, these parameters, above all, depend on the origin of the component assimilated by *Aspergillus niger* and remain constant in a broad range of its concentration in the varnished fabric.

Analysis of the values of parameters obtained gives a possibility to select studied materials in accordance with decrease of the lag-phase duration (Table 5.3) and increase of all or some other growth parameters, i.e. in the sequence due to increase of the material capability to provide growth of the microscopic fungus. It has been found that such sequence is completely appropriate to that obtained on the basis of comparison of standard water absorption of polymers and relative content of organic alimentation for microorganisms in them (Table 5.3). E.g. the higher is standard water absorption (water absorbing capacity) of materials and concentration of carbonic alimentation in them, the greater are parameters m_∞, b, a, and m_0 and the lower is the value of L.

Table 5.3

Parameters of *Aspergillus niger* growth on polymeric materials. Temperature - +29°C, humidity – 98%.

Material	Standard water absorption (Q), %	$m_\infty \cdot 10^3$, mg/cm^2	$b \cdot 10^2$, hour^{-1}	a	$m_0 \cdot 10^3$, mg/cm^2	L, hour
Polyethylene	0.03	0.27	1.5	12.5	0.02	288
Varnished fabric, modified[5]	0.1	0.32	1.8	5.4	0.05	240
Polyvinylchloride	0.2	0.96	1.8	9.6	0.09	216
Poly(methyl methacrylate)	0.4	1.9	2.1	18.0	0.1	168
Varnished fabric	2.0	2,500	2.2	34.7	70	96
Varnished fabric-1 (thermooxidized during 20 hours)	2.0	5,000	2.2	70.4	70	96
Varnished fabric-2 (thermooxidized during 40 hours)	2.3	7,100	2.3	100.4	70	96

These data correlate well with the well-known microbiological notions [132, 133] and data of numerous investigations on microbiological resistance of polymers [45, 275 – 277], which indicate that the increase of water and carbonic alimentation component concentration, accessible for microorganisms, in nutritious media (materials) induces intensification of their growth.

Biochemical properties of microorganisms have been characterized by activity of exoenzymes, produced by them, which are responsible for microbiological growth on polymeric materials [2, 45, 275]. According to ref. [277], microscopic fungi-destructors of materials form the following sequence in the order of reducing activity of such enzymes:

Aspergillus niger > *Penicillium cyclopium* > *Penicillium chrysogenum* > *Paecilomyces varioti*

Table 5.4 shows parameters of these microscopic fungi, obtained by the authors for the biomass growth on varnished fabric. Based on values of the

[5] Varnished fabric was modified by the method of plasma-chemical treatment.

parameters, micromycetes are placed in the Table in descending order of their ability for growth on this material.

Table 5.4

Growth parameters of microscopic fungi on varnished fabric. Temperature - +29°C, humidity – 98%.

Microscopic fungus	$m_\infty \cdot 10^3$, mg/cm^2	$b \cdot 10^2$, hour^{-1}	a	$m_0 \cdot 10^3$, mg/cm^2	L, hour
Aspergillus niger	2.5	2.2	34.7	0.07	96
Penicillium cyclopium	2.3	2.1	37.3	0.6	96
Penicillium chrysogenum	2.2	1.9	30.4	0.07	120
Paecilomyces varioti	0.9	2.0	21.5	0.04	144

Clearly this sequence (Table 5.4) coincides well with the one displayed in ref. [277] for comparative activity of fungi as producers of enzymatic complexes. From this it follows that the higher is activity of enzymes synthesized by a fungus, the more intensively it grows on the varnished fabric. This correlates well with the widespread opinion [2, 45, 275, 277] that the enzymatic mechanism plays the leading role in degradation and utilization of polymeric materials by micromycetes.

Hence, estimation results of materials' and microorganisms' ability to provide for microbiological growth, based on considered growth parameters, coincide with the ones obtained on the basis of the known notions about the effect of participating objects' properties on this process. As a consequence, the parameters suggested adequately characterize growth of microorganisms on the materials.

Clearly quantitative relations between the parameters of microbiological growth and concentration of water and carbonic alimentation in the material, and enzymatic activity of microorganisms can be obtained, for example, the ones analogous to the correlation dependence between Q and adhesion indices of spores to surfaces (Chapter 4). Such relations will allow comparative and forecasting estimations of the ability of microorganisms to grow on technical materials.

Possibilities of applying suggested parameters to development of forecast methods of microbiological degradation of materials under natural conditions have also been estimated. Dependencies of the growth parameters of microorganisms on materials on temperature and air humidity have been studied. The results obtained for *Aspergillus niger*-varnished fabric and *Aspergillus niger*-polyvinylchloride couples are shown in Figure 5.2.

Clearly the growth parameters are sensitive to variation of the hydrothermal mode of tests. Hence, temperature and humidity effect on parameters of the biomass growth on varnished fabric and polyvinylchloride are of the same type. Parameters b, a, m_0 and m_∞ increase and *Aspergillus niger* adaptation to materials shortens with temperature (at constant humidity). Humidity increase (at constant temperature of tests) also induces shortening of the lag-phase and increase of other parameters. Maximal values of b, a, m_0 and m_∞ parameters and the shortest lag-phase have been observed at +29°C and 100% humidity, i.e. under the most favorable conditions for development of microscopic fungi – the destructors of materials.

The dependencies shown in Figure 5.2 testify about different effects of temperature and humidity on every growth parameter. A new parameter Δ representing relative alternation of the growth parameter (Y) at temperature (T) increase from 15 to 29°C (at constant air humidity, φ) or at alternation of φ from 75 to 100% (at constant temperature of the tests) was used to characterize this effect. Figure 5.2 shows also the obtained values of Δ.

It has been found that Δ values calculated for every dependence $Y = f(T)$ (obtained for different air humidity, but constant in the current experiment, φ = const), are almost the same. Values of Δ calculated for a series of dependencies $Y = f(\varphi)$ (obtained at different temperatures, but constant in every particular experiment, T = const), differ also insignificantly. Consequently, investigations of the current temperature-humidity interval have shown that variations of fungus growth parameters on materials are independent of the air humidity at temperature variation and vice versa, it is independent of the test temperature mode under humidity variations. These data allow application of Δ parameter to comparative estimation of the influence of each of these two factors on growth of microorganisms.

Analysis of data obtained on Δ (Figure 5.2) shows that humidity increases from 75 to 100% with a and m_∞ parameters increasing by 25 – 38%. Hence, b, m_0 and L parameters are changed by 3 – 10%, which does not exceed the mistake of their experimental determination. The effect of temperature on the growth parameters is not so high. For *Aspergillus niger* growth on varnished fabric, they are relatively changed from 18 to 76%, and for polyvinylchloride from 24 to 82% (at temperature variation within the range of 15 - 29°C). It is also important that the influence of hydrothermal conditions on fungus growth is higher for cultivation on polyvinylchloride than for varnished fabric. At the same time, it has been already shown (Table 5.3) that varnished fabric possesses higher (compared with polyvinylchloride) capability for providing growth of microorganisms. One may suggest that the less favorable is material for growth of microorganisms, the more important source of substance necessary for growth (water, mineral components) is the environment. As a consequence, influence of the environment on *Aspergillus niger* growth also increases.

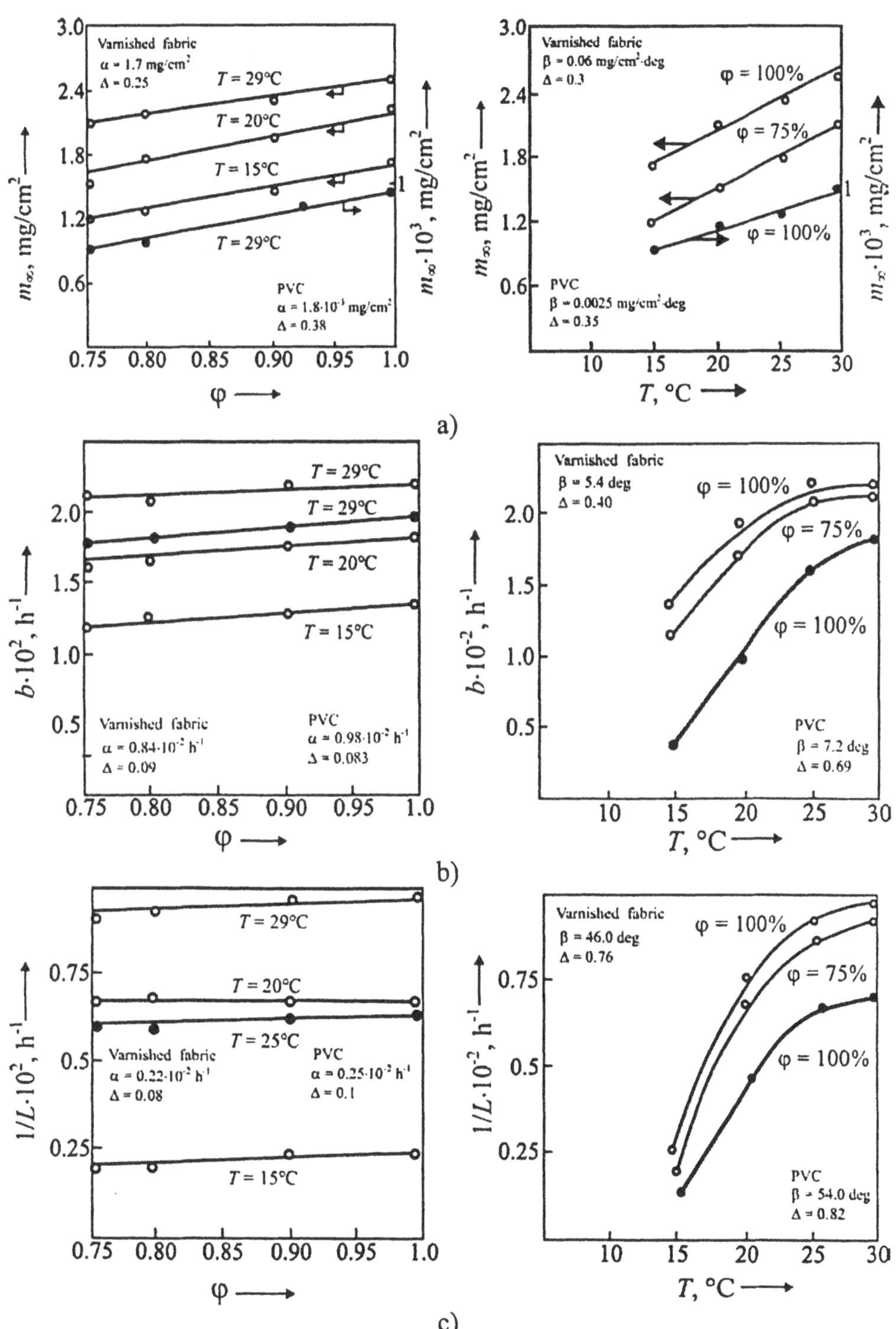
Varnished fabric
α = 1.7 mg/cm²
Δ = 0.25
T = 29°C
T = 20°C
T = 15°C
T = 29°C
PVC
α = 1.8·10⁻³ mg/cm²
Δ = 0.38
m∞, mg/cm²
m∞·10³, mg/cm²
φ
Varnished fabric
β = 0.06 mg/cm²·deg
Δ = 0.3
φ = 100%
φ = 75%
φ = 100%
PVC
β = 0.0025 mg/cm²·deg
Δ = 0.35
T, °C
a)
T = 29°C
T = 29°C
T = 20°C
T = 15°C
Varnished fabric
α = 0.84·10⁻² h⁻¹
Δ = 0.09
PVC
α = 0.98·10⁻² h⁻¹
Δ = 0.083
b·10², h⁻¹
Varnished fabric
β = 5.4 deg
Δ = 0.40
φ = 100%
φ = 75%
φ = 100%
PVC
β = 7.2 deg
Δ = 0.69
b·10⁻², h⁻¹
b)
T = 29°C
T = 20°C
T = 25°C
T = 15°C
Varnished fabric
α = 0.22·10⁻² h⁻¹
Δ = 0.08
PVC
α = 0.25·10⁻² h⁻¹
Δ = 0.1
1/L·10², h⁻¹
Varnished fabric
β = 46.0 deg
Δ = 0.76
φ = 100%
φ = 75%
φ = 100%
PVC
β = 54.0 deg
Δ = 0.82
1/L·10⁻², h⁻¹
c)

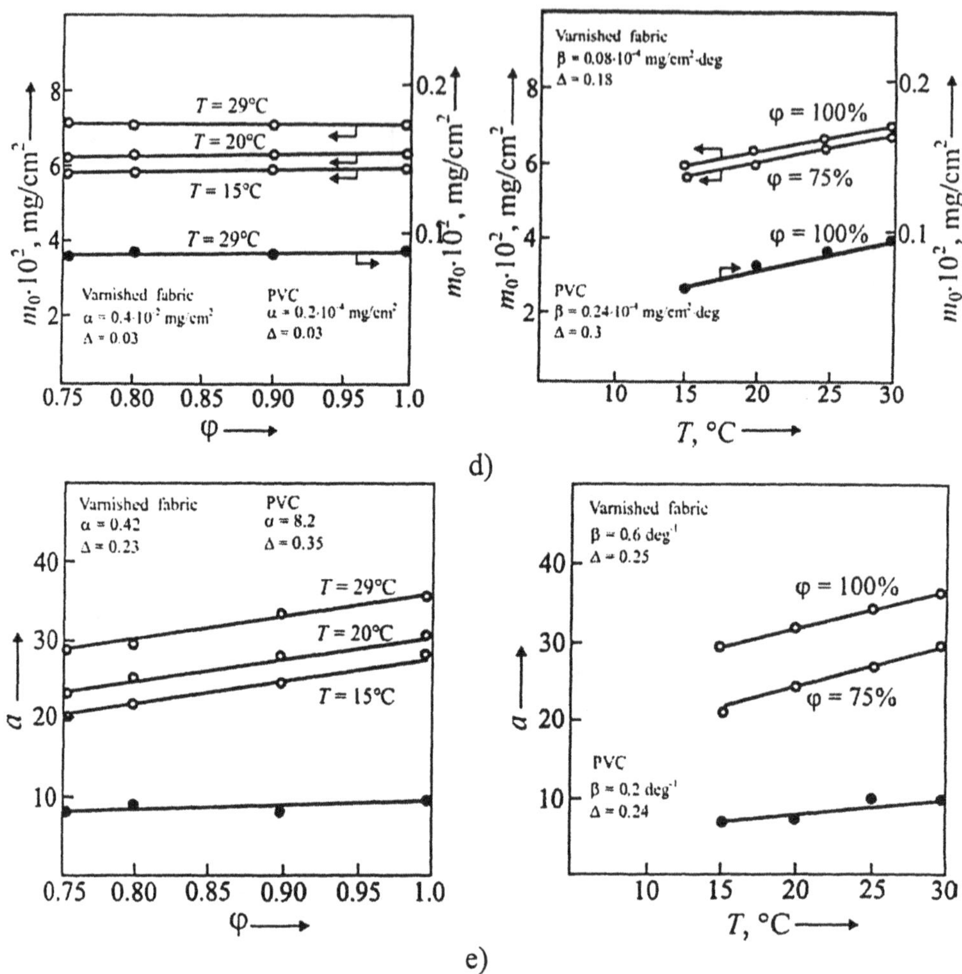

Figure 5.2. Dependencies of m_∞ (a), b (b), L (c) and a (d) parameters on temperature (T) and humidity (φ) of *Aspergillus niger* cultivation on varnished fabric (○) and polyvinylchloride (•)

Hence, m_∞, b, L and m_0 parameters gave an opportunity to estimate quantitatively the effect of temperature-humidity conditions of the environment on the microscopic fungus growth on materials and clear out some features of this process. It has been found that in the temperature-humidity range studied, the determining effect on microbiological development is induced by temperature. Hence, the more favorable nutritious medium the material is for microorganisms, the higher is the effect of hydrothermal conditions of the environment on the growth.

Mathematical treatment of experimental data shown in Figure 5.2 gives analytical correlation expressions (Table 5.5), which bind values of growth parameters with the test temperature-humidity conditions. Dependencies of the growth parameters on air humidity are approximated by equation (5.8). Temperature dependencies of m_∞, a and m_0 parameters are approximated by linear equation (5.9), and b and L parameters by exponential equation (5.10).

In expressions (5.8) – (5.10), humidity (α_Y) and temperature (β^*_n and β_n) coefficients reflect the type and degree of hydrothermal conditions' effect on the appropriate growth parameters.

Calculations have shown that as well as Δ parameter, the values of humidity coefficients are independent of temperature, and vice versa, temperature coefficients are independent of air humidity (Figure 5.2). This gives an opportunity to apply equations (5.8) – (5.10) and obtained values of α_Y, β^*_n and β_n parameters (Figure 5.2) to calculations of the growth parameters at any combination of temperature and humidity in the studied range of their variations.

Table 5.5

Analytical dependencies of *Aspergillus niger* growth parameters on varnished fabric and polyvinylchloride on temperature and humidity

Type of dependence	Propagation parameters	Type of equation
$Y=f(\varphi)$	b, m_∞, a, m_0, $1/L$	$Y^{T_1,\varphi} = Y^{T_1,\varphi_1} + \alpha_n(\varphi - \varphi_1)$ (5.8)
$Y=f(T)$	m_∞, a, m_0	$Y^{T,\varphi_1} = Y^{T_1,\varphi_1} + \beta^*_n(T - T_1)$ (5.9)
	b, $1/L$	$Y^{T,\varphi_1} = Y^{T_m,\varphi_1} \exp\left(-\dfrac{\beta_n}{T}\right)$ (5.10)

Note: T, φ, T_1, and φ_1 are variable and fixed values of temperature and air humidity, respectively; T_m is the maximal temperature (+29°C) used in the experiment; $Y^{T,\varphi}$ is the value of the growth parameter at appropriate indexation of temperature and humidity; α_Y is the humidity coefficient; β^*_n and β_n are temperature coefficients.

Hence, investigation results shown give grounds to conclude that suggested parameters provide for an opportunity to clear out quantitative interconnections between microbiological growth parameters and characteristics of the material, microorganism, and temperature-humidity complex of the environment. This testifies about quite broad possibilities of application of these parameters to comparative estimations of abilities of various microorganism-material couples to provide for growth and gives bases to development of various methods of its forecasting.

Refs. [273, 278] give the experimental proof that thermal inactivation of cells and growth of some species of malignant tumors submit to kinetic regularities of proceeding of the first order chemical reactions. This allowed the authors of refs. [273, 278] the use of known notions, typical of kinetics of simple chemical reactions, in consideration and analysis of the most complicated system of metabolic biochemical processes proceeding in the living organism.

It seemed desirable to estimate the possibility and borders of application of such regularities to description of microorganisms-destructors' growth on materials.

It is commonly known that the rate and the constant of chemical reactions (of the first order) are associated with the Arrhenius temperature equation [279]. Experimental dependence (5.10) can be easily reduced to the form of the Arrhenius equation:

$$b = A_b \exp\left(-\frac{E_b}{RT}\right); \quad (5.11)$$

$$\frac{1}{L} = A_L \exp\left(-\frac{E_L}{RT}\right); \quad (5.12)$$

where A_b and A_L are the pre-exponential coefficients; E_b and E_L are the effective activation energies of the processes; R is the gas constant.

Equation (5.11) reflects temperature dependence of parameter b, which characterizes the specific rate of microorganism biomass growth on the material. The meaning of equation (5.12) can be explained on the basis of the following approximation. One may assume that the lag-phase (L) is ended after accumulation of some quantity of a definite biologically active metabolite, necessary for providing biomass growth, in microbial cells. Then $1/L$ value will characterize the rate of increase of this substance content in cells, and equation (5.12) displays its dependence on temperature.

Table 5.6

Parameters of equations (5.11) and (5.12) for *Aspergillus niger* growth on materials

Material	E_b, kcal	A_b, s^{-1}	E_L, kcal	A_L, s^{-1}
Varnished fabric	13.3	$3.2 \cdot 10^{11}$	39.7	$4.4 \cdot 10^{28}$
Polyvinylchloride	19.8	$1.2 \cdot 10^{11}$	42.2	$2.7 \cdot 10^{28}$

In analog to simple first order chemical reactions, one may suppose that E_b and E_L parameters from equations (5.11) and (5.12) characterize energetic and A_b and A_L parameters spatial hindrances of the processes determining rates of microorganism adaptation to the material and growth of the biomass, respectively. Table 5.6 shows values of these parameters for *Aspergillus niger* growth on varnished fabric and polyvinylchloride.

Obtained values of parameters are comparable with activation energies and pre-exponential coefficients typical of the first order chemical reactions, which equal 25 – 60 kcal and 10^{10} – 10^{16} s^{-1}, respectively, for such reactions [278, 280].

At the same time, of special attention are anomalously high values of A_L parameter (~10^{28} s^{-1}) comparing with the chemical reactions. Analogous discrepancy was observed in ref. [278] in studies of thermal inactivation of microbial cells. Pre-exponential coefficients in the Arrhenius equation, obtained for this process, equaled 10^{20} – 10^{40} s^{-1}. This is explained by specificity of biochemical break of many (not a single) chemical bonds at thermal degradation of albuminous compounds. Nevertheless, it is shown [278] that the Arrhenius equation displays well the main regularities of the biological processes studied.

The data from Table 5.6 show that values of equation (5.11) parameters differ from corresponding parameters of equation (5.12) for both varnished fabric and polyvinylchloride. One may suggest that the rates of microorganism development at stages of its adaptation to material and biomass growth are controlled by various biochemical processes. This hypothesis correlates well with the data [132, 133] displaying that adaptation of microbial cells to a nutritious substrate is associated with proceeding of biochemical synthesis of enzymes and other biologically active substances, necessary for vital activity of the microorganism on the current substrate, specific for this phase only.

Values of E_L and A_L are higher than corresponded E_b and A_b values. As a consequence, the processes which control duration of biodestructor adaptation to the material require higher energy but are less spatially hindered comparing with the ones limiting the biomass growth rate after the lag-phase termination.

Values of A_b and A_L calculated on the basis of data on tested varnished fabric are 1.5-3-fold greater that analogous parameters of *Aspergillus niger* growth on polyvinylchloride. On the contrary, E_b and E_L parameters for varnished fabric are lower than for polyvinylchloride. This means that, in generally, growth of the microscopic fungus on varnished fabric is less energetically and spatially hindered than on polyvinylchloride. This testifies about higher ability of varnished fabric (rather than polyvinylchloride) to provide for growth, which completely correlates with the above-discussed data. Hence, analysis of obtained values of parameters from equations (5.11) and (5.12) proves the suggestion about analogy between regularities of the first order chemical reactions and growth of the microorganism biomass on technical materials.

At the same time, one should note that the application of chemical kinetics ideas of simple chemical reactions to analysis of microbiological development is sometimes in a definite contradiction to the experimental data obtained.

For example, it is common knowledge [279, 280] that mentioned chemical reactions display variations of one or another controlled parameter (Y) obeying equation of the following type:

$$Y = Y_\infty[1 - \exp(-kt)], \qquad (5.13)$$

where k is the rate constant of the chemical reaction.

Obtained dependencies, $m_t = f(t)$ (Figure 5.1), are S-shaped and most accurately approximated by expression (5.4).

Of special interest is determination of the applicability limits of simple chemical reactions' regularities to description of the microorganism growth. For this purpose, experimental dependencies, $m_t = f(t)$, have been approximated by kinetic law (5.13), which represents as follows in this case:

$$m_t = m_\infty\{1 - \exp[k_m(t - L)]\}, \qquad (5.14)$$

where k_m is the rate constant of the biomass growth.

It has been found that $m_t - t$ dependencies, represented in coordinates of linearized equation (5.14), are curvilinear at the initial period of the biomass growth (Figure 5.3). Obviously, these periods (corresponded to curvilinear parts) neighbor the end of accelerated irregular growth of biomass. After that $\ln\left(1 - \frac{m}{m_\infty}\right)$ - $(t - L)$ dependencies become of the linear type, which is preserved during the rest time of growth proceeding (Figure 5.3). Hence, after the end of irregular accelerated increase of the biomass, the part of the growth curve is successfully approximated by equation (5.14).

Obviously, violation of the linear type of considered graphs is associated with the difference in processes controlling microscopic fungus development at the stages of its adaptation to material and biomass growth. In this case, accelerated irregular growth can be considered as somewhat a transitional period. The rate of biomass accumulation during this period is determined by permanently changing ratio of rates of the processes limiting microorganism development in the lag-phase and the exponential phase. Naturally, the part of biomass growth kinetic curve corresponded to such transitional period cannot be described by equation (5.14) with a constant value of the rate constant, k_m, which leads to violation of linearity of graphic dependencies shown in Figure 5.3.

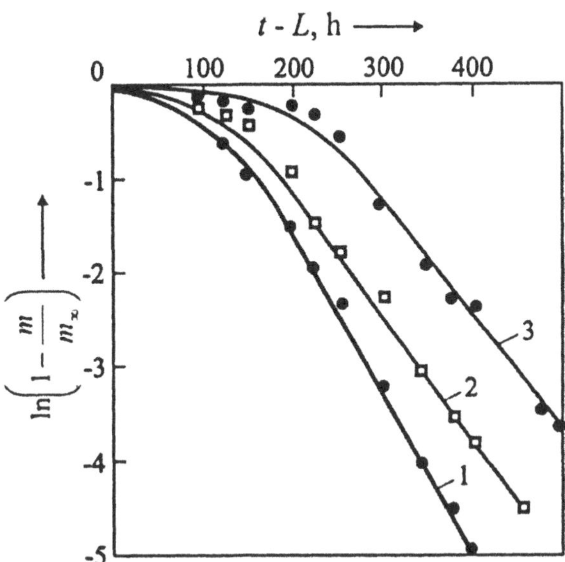

Figure 5.3. Dependence of the biomass quantity (*m*) of *Aspergillus niger* on time of cultivation (*t* – *L*) on varnished fabric (•) and PVC plasticate (□) at 100% humidity and 29°C (1, 2) and 15°C (3)

Hence, the analysis shows that adaptation of a microorganism to material and biomass growth (after the end of the stage of accelerated irregular increase) proceeds due to kinetic regularities of the first order chemical reaction. Temperature dependencies of the lag-phase duration and the specific rate of microorganism biomass growth on materials obey the Arrhenius equation.

The data obtained show availability of the technique and ideas typical of chemical kinetics to be used for investigations of microorganisms' growth on technical materials (its estimation and forecasting). Application of this methodology can be based on suggested experimental technique for determination of parameters and analytical apparatus of quantitative indices of the microbiological growth.

Detected regularities bring out clearly the abilities of the growth in studies of some known [275] methods of its control. For example, data shown in Tables 5.3 and 5.4 and Figure 5.2 experimentally prove suitability of the growth parameters to estimation of efficiency of the following methods: variation of water absorption (hydrophilic-hydrophobic balance) of the material and content of components in it, consumed by microorganisms as nutritious elements; variation of temperature-humidity and other conditions of the environment.

However, carried out investigations have given no proof of parameters' sensitivity to the ability of various chemical substances to inhibit growth of microorganisms-destructors. At the same time, modification of materials by substances possessing this ability (biocides) is the most widespread protection method against microbiological damaging.

The influence of substances possessing biocidal activity on *Aspergillus niger* growth kinetics has been studied. Concentrations of biocides below recommended level for full suppression of microbial growth have been used. Dry specific biomass of the fungus, determined by gravimetric method, was accepted as the growth parameter. *Aspergillus niger* was cultivated on a hydrogel support contacting with a liquid nutritious medium and biocidal substance, specially suggested by the authors of the present monograph [281, 282]. Hydrogel represents a 3D-crosslinked poly(hydroxyethyl methacrylate) possessing a porous structure in block. Application of such support is stipulated by test results on the biocidal activity of various compounds using agar (solid) nutritious media, widely used for this purpose. It has been found that water-insoluble biocides are distributed irregularly by the body and the surface of the medium, which gives no opportunity to estimate their activity. Application of the hydrogel allows elimination of this disadvantage. Distribution regularity of water-insoluble substance is obtained by applying them on such support from solution in an organic solvent. Water-soluble compounds are injected directly to the liquid nutritious medium contacting with the hydrogel. Liquid nutritious medium penetrates freely into the microorganism through the hydrogel porous body.

It has been found that all experimental kinetic curves of *Aspergillus niger* biomass growth on a hydrogel support in the presence of various concentrations of both water-soluble and insoluble biocides are described well by equation (5.4). Figure 5.4 shows generalized kinetic curve of the biomass growth, obtained using the experimental data[6].

Table 5.7 shows values of equation (5.4) parameters for microorganisms' growth on a hydrogel support in the presence of studied biocides.

Analysis of the data shown indicates that common regularities of the substance effect on microscopic fungus growth parameters on hydrogel are observed independently of the type of the substance studied. Initial biomass, m_0, and degree of its growth, a, are practically independent of the type of biocide

---

[6] Generalized equation of growth is as follows: $m/m_\infty = 1/[1 + \ln a - b(t - L)]$. It is obtained by dividing expression (5.4) by m_∞ and bringing in a parameter to the exponent.

and its concentration. Analysis of m_∞ values indicates just weakly expressed tendency to their reduction with the increase of tested substance concentration. Parameters *b* and *L* are much more sensitive. As would be expected, *b* parameter is reduced and duration of fungus adaptation to nutritious substrate increases with the biocide concentration.

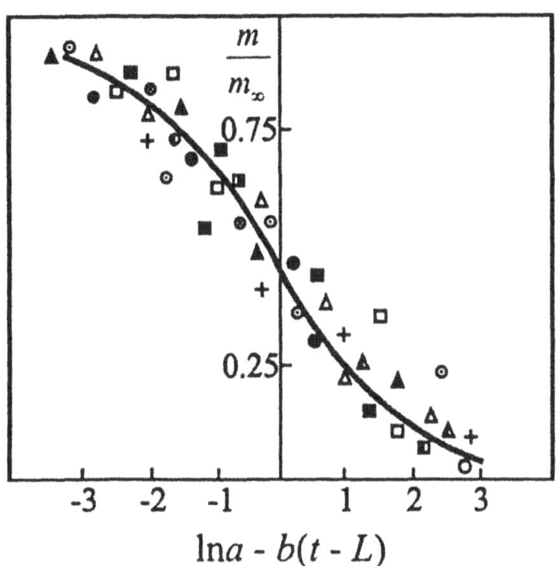

Figure 5.4. Generalized kinetic curve of *Aspergillus niger* biomass growth in the presence of various concentrations of biocides: nicthedin (O), ABDM (Δ), ionol (+), flamal (◉), merthiolate (□), $CuSO_4$ (•), ODP (■), PTMI (▲), pentachlorophenol (▲), salicylanilide (◐), trilan (◧), biocine (⊗)

The data shown in Table 5.7 indicate *b* parameter reduction by 13% and *L* parameter increase by 20% in the presence of 0.001 mg/cm^2 concentration of ODP biocide. Similar effect on these parameters is caused in the presence of 0.1 mg/cm^2 flamal concentration. This means that flamal possesses lower biocidal ability, than ODP. Consequently, parameters *b* (specific rate of the biomass growth) and *L* (duration of the microorganism adaptation to the material) allow quantitative characterization of the growth features in the presence of biocides and estimation of biocidal activity of various substances.

Suppression of the microbial growth by various substances is often associated with their inhibition by enzymatic biochemical reactions in cells of microorganisms [159, 160]. This gives grounds to apply one-factor expression

(5.15) of the uncompetitive deceleration of enzymatic reaction type equation, shown in Table 5.8, to analysis of the experimental data obtained [159, 160].

Table 5.7

Parameters of equation (5.4) for *Aspergillus niger* growth on a hydrogel support in the presence of biocides

Biocide concentration, mg/l (mg/cm^2)	$m_0 \cdot 10^2$, mg/cm^2	a	m_∞, mg/cm^2	b, hour^{-1}	L, hour
Water-soluble substances					
Merthialate					
0.1	3.5	108	3.82	3.3	72
0.3	3.4	110	3.81	2.6	408
0.4	3.4	105	3.64	2.5	624
Nicthedin					
30	3.5	110	4.02	2.8	96
50	3.6	105	3.84	2.2	168
70	3.7	100	3.75	2.1	408
Water-insoluble substances					
ODP					
0.001	3.6	105	3.81	3.1	48
0.003	3.6	97	3.58	2.1	72
0.05	3.5	82	2.87	1.7	168
Flamal					
0.10	3.7	100	3.72	3.2	48
0.25	3.6	113	4.03	2.7	72
0.5	3.6	110	3.71	2.1	120

Note: Parameters of equation (5.4) for *Aspergillus niger* biomass growth on a hydrogel support contacting with Chapek-Dox liquid nutritious medium are the following: $m_0 = 3.5 \cdot 10^{-2}$ mg/cm^2, $m_\infty = 4.0$ mg/cm^2, $a = 113$, $b = 3.6 \cdot 10^{-2}$ hour^{-1}, $L = 40$ hours.

Table 5.8

Analytical dependencies of *Aspergillus niger* growth parameters in the presence of biocides on biocide concentration

Growth parameter	Type of equation	
b	$b_c = b_0 K_b/(K_b + C)$	(5.15)
L	$L_c = L_0 \exp(K_L C)$	(5.16)

Note: b_c, b_0, L_c, and L_0 are values of b and L parameters of the biomass growth in the presence and in the absence of a biocide in concentration C, respectively; K_b is the constant numerically equal to the biocide concentration, at which $b_c = b_0/2$; K_L is a constant.

It has been found that obtained values of b parameter are successfully approximated by equation (5.15). In this case, the correlation parameter equals, at least, 0.87. Calculated values of K_b constant are shown in Table 5.9. Clearly a definite value of this constant corresponds to every biocide. This constant is independent of its concentration and characterizes ability of the substance to inhibit growth of the microorganism. The higher K_b is, the higher concentration of appropriate biocide should be used to decelerate microbial growth by one and the same value, i.e. the lower is the biocide efficiency. For example, $K_b = 0.76$ mg/l for merthiolate and 80.5 mg/l for nicthedin, i.e. merthiolate is by two orders of magnitude more active than nicthedin. Hence, the lower is K_b, the more efficient biocide is.

It has bee found that for all studied substances, dependence of the lag-phase duration on concentration of biocides is described by exponential equation (5.16) from Table 5.8.

Table 5.9

Values of K_b and K_L constants for studied biocides

Biocide	Constants	
	K_b, mg/l (mg/cm^2)	K_L, l/mg (cm^2/mg)
Water-soluble substances		
Merthiolate (sodium ethyl mercuric thiosalicylate)	0.8	6.7
ABDM (alkyl benzyl methyl ammonium chloride)	8.9	0.2
Nicthedin (1,6-diguanidine hexadihydrochloride)	80.5	0.03
$CuSO_4$ (copper sulfate)	1,952	0.0005
Water-insoluble substances		
ODP (o-phanylphenol)	0.004	250
PTMI (n-paratolylmaleimide)	0.05	62.5
Flamal (bis(0,0-1-chlorotribromisopropyl)-3-chlor-2-bromopropylphosphonate)	0.7	1.9
Ionol (2,6-ditertbutyl-4-methylphenol)	17.5	0.15
Pentachlorophenol	0.01	125
Salicylanilide	3.8	44.5
Trilan (4,5,6-trichlorbenzoxazolone-2)	0.07	21.2
Biocin (double zinc salt of ethylene-bis-dithiocarbamic acid and methyl ester of N-benzimidazolyl-2-carbamic acid)	10.0	5.2

Table 5.9 shows also values of K_L constant from this equation for studied substances. This constant represents one more parameter of biocidal properties (for the given substance), independent of the concentration. The higher is K_L, the longer is the lag-phase at one and the same biocide concentration, i.e. the more efficient the biocide is.

It has been found that K_b and K_L constants are uniquely bind to one another and to biocidal activity of the substance. Values of the constants calculated from the experimental data obey the following linear dependence:

$\frac{1}{\ln K_b} = f(\ln K_L)$ (Figure 5.5). This dependence gives an opportunity to make a preliminary estimation of the biocidal properties of various compounds and their concentration in materials, necessary for reliable protection from microbiological damaging.

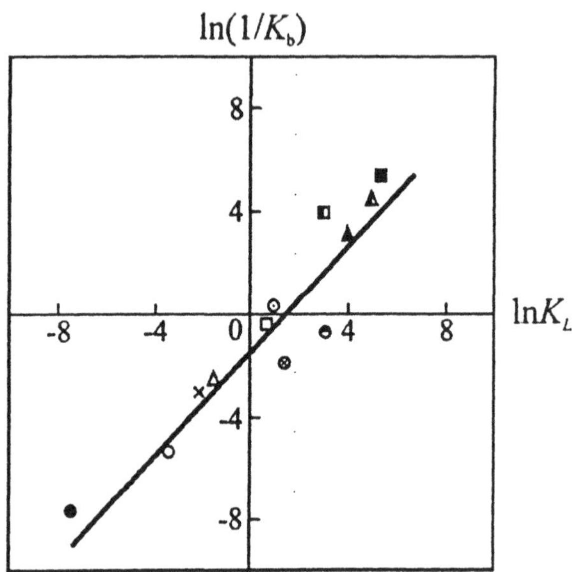

Figure 5.5. Generalized dependence binding biocidal activity constants, K_b and K_L, for various substances (refer to Figure 5.4 for notations)

Hence, suggested parameters can be used for selection, efficiency estimation and creation of particular methods and means of the microorganism growth control on materials.

CONCLUSION

As indicated, the biomass growth of microscopic fungi and bacteria on all studied materials obeys regularities peculiar to growth of microorganisms in a periodical culture. The growth is characterized by several stages changing one another in the definite sequence as follows: the lag-phase, exponential and stationary stages. Durations of stages are determined by the material origin, properties of microorganisms and environmental features.

Kinetics of all the stages of microorganism biomass growth on materials obeys common exponential equation – the growth equation. Parameters of the growth equation consider and characterize growth features for various material-microorganism couples and can be used as quantitative parameters of the current stage of microbiological damaging. Such parameters are as follows: maximal biomass reached in the experiment (m_∞); duration of microorganism adaptation to material (the lag-phase) (L); specific rate of the biomass growth (b).

The method for determination of these parameters considering calculation of their values based on kinetic dependencies of biodestructor biomass variation during cultivation on the material, obtained in the experiment, has been proved experimentally.

Suitability of the growth parameters and the technique for their determination to estimate and forecast microbiological resistance of materials and efficiency of protection measures have been shown.

Interconnection between standard water absorption of polymeric materials, activity of enzymes produced by microorganisms and growth parameters has been determined quantitatively. An increase of water absorption of the materials and enzymatic activity of fungi is accompanied by shortening of the lag-phase (L), increase of the growth rate (b) and maximal biomass value (m_∞).

It has been shown that the period of *Aspergillus niger* adaptation to polyvinylchloride and varnished fabric, as well as the rate of biomass growth on these materials in the studied temperature-humidity interval (t = +15 – +29°C, φ = 75 – 100%) are mostly defined by temperature. Influence of the humidity mode on these parameters is insignificant. Analytical dependencies binding the growth parameters with temperature and environmental humidity have been obtained.

It has been found that the whole cycle of microorganism biomass growth on material, except for an insignificant period of its accelerated irregular increase, proceeds due to kinetic regularities of the first order chemical reactions. Temperature dependencies of the microorganism adaptation duration to material and specific rate of the biomass growth obey the Arrhenius equation.

It has been shown that substances possessing biocidal properties induce the appropriate increase of the period of microorganism adaptation to nutritious medium (the lag phase – L) and reduction of the biomass growth rate (b). Analytical equations binding these parameters with concentration of biocides have been suggested. The technique for quantitative estimation of biocidal properties of the substances providing for application of hydrogel support to microorganism cultivation, obtaining a series of kinetic dependencies of growth at different concentrations of tested compound, calculation of L and b growth parameters and then K_b and K_L constants, which characterize the biocidal activity, have been developed.

Chapter 6.

Material properties variation under the action of microorganisms

Changes in properties of materials contacting with microorganisms-destructors is considered by the authors as the third (after adhesion and growth of the mass of microbial cells) stage of microbiological damaging proceeding in parallel with the biomass growth and frequently proceeding after its end. In the majority of cases, this very stage stipulates occurrence of breakdowns and disorders of technical articles, which is associated with a biofactor.

Experimental results of reproduction of the real microbiological damage type (Chapter 3) have given grounds to a supposition that variation of a series of material properties important for workability of technique under the effect of biodestructors proceeds mostly due to the effect of specific liquid aggressive media – aqueous solutions of substances excreted by microorganisms (metabolites). Therewith, in this case, deterioration of the properties can be caused by both physical (sorption, desorption, loading by biomass) and chemical (chemical and electrochemical transformations) of interactions between materials and metabolites.

This Chapter gives the experimental proof of hypotheses on the origin of the microorganism effect on the properties of materials. Kinetic equations and parameters of variation in properties of materials interacting with biodestructors have been analytically deduced; methodical approach to the study of this stage of microbiological damaging as well as the possibility of using such approach to estimation and forecasting of the process and efficiency of protection measures have been proved.

6.1. CHANGES IN MATERIAL PROPERTIES INDUCED BY PHYSICAL PROCESSES

Analysis of the experimental results on reproduction of the type of real microbiological damages of components and associates of the articles (Chapter

3) as well as analysis of modern ideas on possible mechanisms of the effect of physically aggressive media on materials have given rise to a supposition that variation of strength and dielectric properties of polymers contacting with microscopic fungi can be stipulated by the main physical processes as follows:

- adsorption of microorganisms' metabolites on the material surface;
- sorption of metabolites in the polymer volume;
- desorption of low-molecular components from the polymeric material.

Microbiological deterioration of fuels and lubricants, such as concentration of mechanical admixtures in them, can be related to naturally physical processes. In this case, its change is the result of formation and accumulation (growth) of the microbiological mass in fuels and lubricants (poisoning of fuels and lubricants by biomass).

Let us consider proofs for the possibility, main regularities and parameters of the mutual influence of the mentioned processes on the material properties.

6.1.1. Adsorption of microscopic fungus metabolites on the surface of polymeric materials

It is common knowledge [283 – 295] that adsorbed molecules of the medium may reduce the phase surface energy at the polymer-medium interface and increase surface conductivity of dielectrics. The first effect makes formation of new surfaces simpler at the material deformation and appropriate variation of its mechanical properties. The second effect induces changes in dielectric parameters.

Experiments on reproduction of the real biodamages have displayed that induced elasticity (σ) of PE, PVC, PMMA, and CTA, durability (τ_d) of PMMA, as well as electrical insulation resistance (R) of PVC-plasticate are reduced already during formation of fungal colonies (biomass growth). Hence, it has been found that observed changes in properties are reversible, i.e. if the biomass is removed from the samples with their further conditioning, the controlled parameters are restored up to the initial values. As is known [171 – 173], such reversibility of property variations proves that the processes stipulating these variations are of the physical origin and, consequently, can be associated with the adsorptive mechanism of the medium effect (metabolites of fungi) on polymers and (or) with its sorption in the material volume.

In another series of experiments the parameters were measured using samples thoroughly cleaned from biomass, but without conditioning. In this case, reduction of the controlled parameters would supposedly be determined by sorption of metabolites in the material volume. It has been found that the effect of biodestructors is significantly (by 3 -6 times) lower than that obtained for the samples loaded with biomass.

These results have given bases to a supposition that adsorptive changes of the surface energy and electric conductivity of polymers contacting with microorganisms is the dominant process inducing changes in their properties.

The results of model experiments determining durability (τ_d) of PMMA contacting with the biomass should be mentioned separately. It has been found that obtained durability curves presented in the Zhurkov equation coordinates ($\lg\tau_d - \sigma_d$) have possessed a linear part at stresses, σ_d, exceeding 21 MPa. It is known [171, 172] that linearity of the mentioned dependence testifies about thermofluctuation type of degradation. Hence, realization of such mechanism under the action of an aggressive medium allows association of changes in τ_d with adsorptive effect of the medium on the stressed material.

Figure 6.1 shows dependencies of strength and dielectric characteristics of polymers, as well as the quantity of *Aspergillus niger* biomass formed on them on the incubation time of samples, poisoned with fungal spores. Parameters of the properties were measured without biomass removing from the sample surface. Durability was determined under the tensile stress (σ_d) equal 25 MPa.

The data from Figure 6.1 show that kinetic curves of PMMA short- and long-term strength variation and PVC dielectric properties are characterized by common shape and similarly relate to the stages of biomass growth on the materials. This may testify about community of processes stipulating the observed changes in properties. At the initial stage of contact with microbial cells, kinetics and the total effect of changes in registered characteristics of materials are practically the same as for the control samples. One may conclude that during this period, changes in properties are not induced by metabolism of microorganisms, but probably by air humidity influence on the materials [296] (experiments were carried out at 100% humidity).

The effect of *Aspergillus niger* is observed only after the end of microscopic fungus adaptation to the material (the lag-phase). Then the biomass growth is accompanied by a monotonous decrease of σ, τ_d and R, which is terminated almost simultaneously for these parameters, when the microorganism reaches the stationary stage of development.

One should note that at contact with the biomass, controlled parameters are changed by 85 – 95% of the total effect of *Aspergillus niger* on the material during the stages of its accelerated irregular and exponential growth. This means that the studied change in properties mostly happens during maximal activity of microorganisms as producers of exometabolites [136, 277].

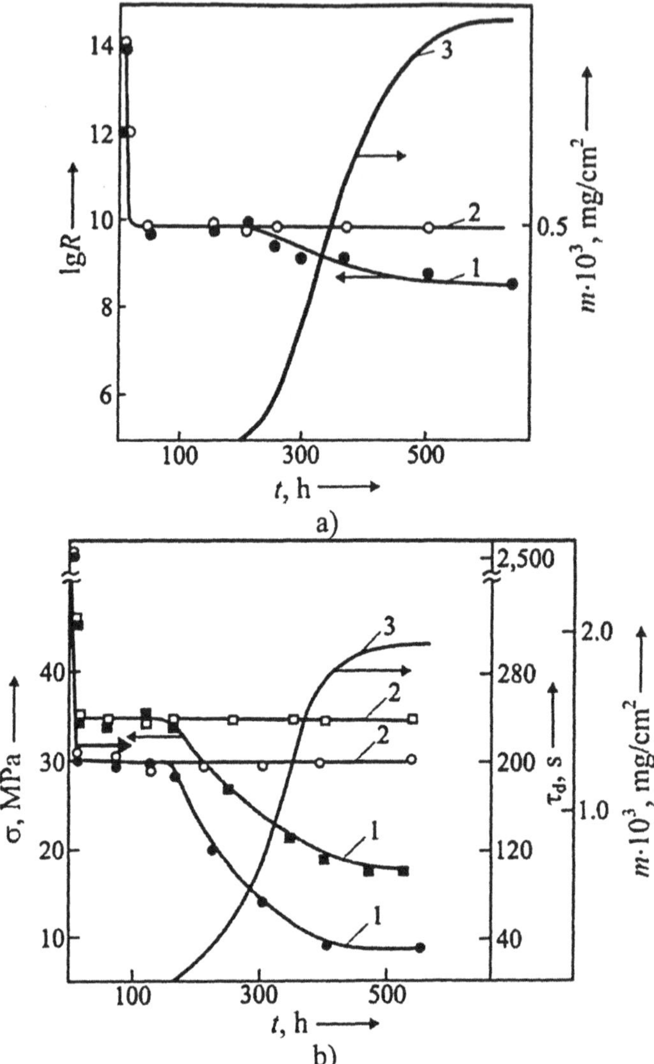

Figure 6.1. Dependencies of changes in: a) electrical resistance (*R*) of PVC; b) induced elasticity limit (σ) and durability (τ_d) of PMMA on time of contact between polymers and *Aspergillus niger*.
1, 2 – changes in properties of samples poisoned by microbial cells (•, ■) and control ones (○, □), respectively; 3 – kinetic curve of *Aspergillus niger* biomass growth on polymers. Mechanical parameters have been determined: σ at the deformation rate equal 0.01 mm/s and τ_d at the stress equal 25 MPa.

A significant increase of surface-active metabolite concentration in solution contacting with the material and, consequently, an increase of their quantity, adsorbed to the material surface, is also the most probable during this very period. If the adsorption hypothesis concerning changes in properties of polymers is true, such increase of the adsorbate surface concentration explains the observed type of reduction of PMMA strength and PVC-plasticate electrical resistance.

Analytical model of the studied process has been obtained on the basis of the Langmuir equation of monomolecular adsorption, which connects substance concentrations in the solution and on the solid surface [289]. Hence, based on the results of multiple investigations of biosynthetic properties of microscopic fungi and bacteria [136, 297, 298], it has been assumed in the first approximation that the equilibrium concentration (C_{pm}) of surface-active metabolites on the polymer surface is linearly dependent on the colony biomass (m):

$$C_{pm} = \chi m,$$

where χ is a coefficient.

In accordance with the adsorption hypothesis, reduction of strength and dielectric properties of the material should be proportional to the change of the phase surface energy at the polymer-solution of metabolites interface and the surface electric conductivity, respectively. In turn, the surface energy and the current conductivity are determined by the quantity of metabolites adsorbed on the material, a_m. In this case, changes in σ, τ_d and R should be proportional to a_m. Using the Langmuir equation for transiting from a_m to equilibrium concentrations of metabolites, C_{pm}, and expressing these concentrations via m, we obtain:

$$\frac{m}{(A_0 - A)/A_0} = \frac{1}{\chi\alpha_n a_p K} + \frac{m}{\chi\alpha_n a_p} = P_A + N_A m, \qquad (6.1)$$

where A and A_0 are values of a polymer property at contact with biomass and in humid atmosphere (control), respectively; α_n is the constant binding the relative change of controlled parameter of the property, $\Delta A/A_0$, and a_m ($\Delta A = A_0 - A$); a_p is the capacity of dense monolayer of the adsorbate on the polymer surface; K is a constant characterizing properties of the surface-active substance (metabolites).

It has been found that the changes in strength and dielectric properties of materials contacting with the biomass at the stages of its accelerated irregular and exponential growth are successfully approximated by equation (6.1). In this case, correlation indices are, at least, 0.9.

As the biomass exceeds a definite value, m_A^*, determined for every material (Table 6.1), which corresponds to transition from the exponential to the stationary stage of *Aspergillus niger* growth, changes in σ, τ_d and R are not yet submit to equation (6.1). Hence, correlation indices do not exceed 0.6.

Figure 6.2 illustrates estimation results of the model (6.1) correspondence to the experimental data. Clearly A - m dependencies presented in the graphic form in coordinates of equation (6.1), $m/(\Delta A/A_0) - m$, display a linear part. At high values of the biomass, m_A^*, graphic dependencies become nonlinear. One may suggest that distortion of linearity is associated with variation of the process type stipulating changes in the properties. Obviously in the case of influence of large biomasses ($m > m_A^*$) on the material and, consequently, solutions with high concentrations of metabolites, reduction of PVC-plasticate electrical resistance of may be caused by formation of both adsorbed and phase film of current-conducting liquid on the dielectric surface.

As mentioned above, besides adsorptive action of metabolites, reversible changes in PMMA properties may be induced by their sorption in the polymer volume. It is common knowledge that one of the typical signs of such mechanism of medium action is changing the initial elasticity modulus of the material contacting with it. The authors of the present monograph have detected experimentally that the initial elasticity modulus of PMMA remains unchanged and does not differ from the one obtained in control tests independently of the biomass volume contacting with PMMA. This circumstance allows a supposition that distortion of linearity of $m/(\Delta\sigma/\sigma_0) - m$ and $m/(\Delta\tau_d/\tau_{d0}) - m$ (at $m > m_A^*$) dependencies is stipulated by preferable penetration of metabolites into local volume of deformed polymer before the tip of propagating crack. This results in plasticization of the material in this volume (the so-called local plasticization) and appropriate change in the crack propagation rate. Local plasticization causes no effect on the initial elasticity modulus, because it is associated with the cracks formed during PMMA stretching at stresses close to the induced elasticity limit only. The features of such mechanism of the metabolite effect on polymer properties are discussed in Section 6.1.2.

Table 6.1 shows values of equation (6.1) parameters calculated for the linear parts of $m/(\Delta A/A_0) - m$ dependencies.

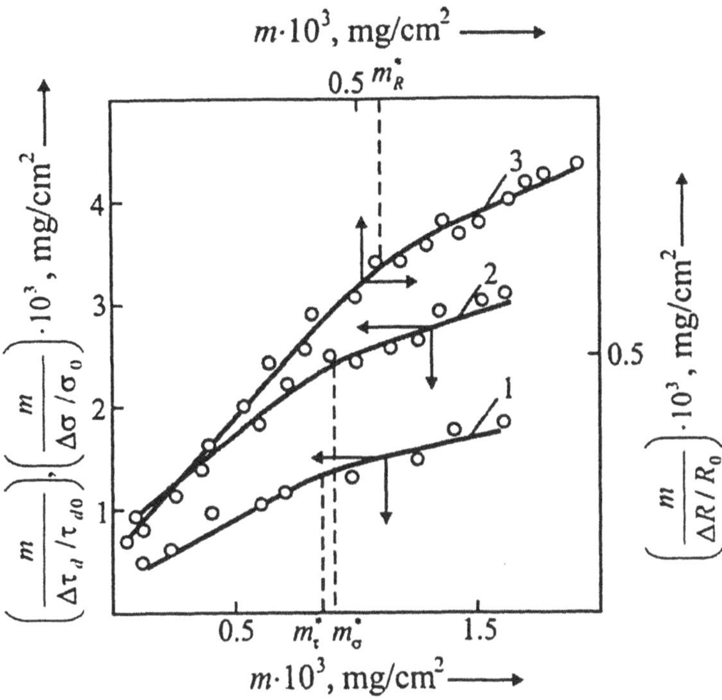

Figure 6.2. Dependencies of: 1 - relative durability change (τ_d), 2 - induced elasticity (σ) of PMMA and 3 – electrical insulation resistance (R) of PVC on *Aspergillus niger* biomass ($R_0 = 6.3 \cdot 10^9$ Ohm; $\sigma_0 = 35$ MPa; $\tau_0 = 200$ s)

Table 6.1

Values of equation (6.1) parameters for variations of polymer properties at the contact with *Aspergillus niger* biomass

Material	Property index	$m^*_A \cdot 10^{-3}$, mg/cm^2	$P_A \cdot 10^{-3}$, mg/cm^2	N_A	$K \cdot 10^3$, cm^2/mg
PVC	R	0.55	0.14	1.31	9.2
PMMA	σ	0.90	0.75	2.52	3.3
	τ_d	0.85	0.35	1.25	3.6

The data from Table 6.2 show that the changes in PVC electric conductivity and short- and long-term strength of PMMA are characterized by a specific, peculiar to them only, selection of P_A and N_A values. Actually, according to equation (6.1), P_A and N_A are inversely proportional to multiplied coefficients, $\sigma_A \times \chi$. Coefficient σ_A binds relative variation of the particular property of the material to the quantity of metabolites absorbed on its surface.

Obviously, the same absorptive surface concentration may cause different effects on each parameter, for example, τ_d and σ of PMMA, and, consequently, α_τ and α_σ coefficients should also be different.

Coefficient χ characterizes the equilibrium concentration of surface-active metabolites in solution contacting with the material during biomass growth. According to the meaning, it depends on the material origin and biosynthetic features of the microorganism. This is the reason why χ and σ_A values should be different for PMMA - *Aspergillus niger* and PVC - *Aspergillus niger* couples.

According to the model (6.1), the coefficient, K, depends on properties of surface-active metabolites. As a consequence, it should be constant for the current microorganism – material couple independently of the property index to be determined. Table 6.1 shows data testifying fulfillment of this condition. Values of K for the *Aspergillus niger* – PMMA couple, calculated from the test data on long-term (τ_d) and short-term (σ) strength, differ insignificantly.

Hence, obtained values of N_A, P_A and K parameters do not contradict to their physical meaning in the analytical model (6.1).

Carried out analysis gives grounds to a conclusion that reduction of mechanical parameters for PMMA and dielectric indices for PVC at their contact with the biomass during its accelerated irregular and exponential growth obeys the model (6.1), analogous to the Langmuir monomolecular adsorption equation. Equation (6.1) sets relation between the biomass quantity present on the material at any moment and the effect of material property variation.

It is common knowledge [289, 299] that adsorption of dissolved substance molecules to a solid surface proceeds almost immediately. Then, in analogue, the adsorptive change in properties of polymers should proceed simultaneously with the biomass growth (metabolites' concentration increase) on the material. In this case, the rate of microbiological deterioration of properties will be determined by the biomass growth rate. Chapter 5 shows that the biomass growth on polymeric materials obeys the exponential equation (5.4). Substituting variable m in the analytical model (6.1) by this equation, we obtain the expression reflecting kinetics of changes in the operational properties of polymers due to adsorptive effect of microorganism metabolites:

$$A_0 - A = \frac{A_0 \cdot m_\infty}{P_A + N_A a \exp[-b(t-L)] + N_A \cdot m_\infty}. \tag{6.2}$$

Carried out calculations have shown high correspondence between expression (6.2) and the experimental data (Figure 6.1). Correlation indices fall within the range of 0.91 – 0.95 due to the type of studied material and controlled property.

The analysis of equations (6.1) and (6.2) allows a suggestion of quantitative indices of the considered stage of microbiological damaging. Parameters P_A and N_A can play the role of such indices. They provide for a possibility of estimating the surface activity of metabolites and variation of the material properties. Biomass growth indices, studied in Chapter 5, reflect the process rate.

The complex of the above-mentioned parameters is sensitive to the features of the adsorptive mechanism of microorganism impact on polymers and practically the most valuable period of the process, when the properties are changed by 85 – 95% of their maximal value obtained during the initial period of contact between the object and the colony formed on it.

The adsorptive origin of changes in properties of materials contacting with a biodestructor was proved in experiments with the model liquids. Such media included separate metabolites, the most aggressive to polymers: single, bi- and tri-basic carboxylic acids [286, 297, 300, 301]. Changes in PMMA properties during its deformation in water solutions of theses acids have been studied (Figures 6.3 and 6.4).

It has been found that succinic, oxalic, fumaric, tartaric and citric acids cause no significant effect on mechanical characteristics of the polymer. On the contrary, monobasic acid solutions seriously change the induced elasticity limit and durability of PMMA.

Analyzing data shown in Figure 6.3, one can easily observe that transiting from acetic to butyric acid, σ values are reduced uniformly in definite concentration ranges, if concentration of every consequent homologue is about 3 times lower than of the previous one. Thus one may suggest that the known Duklo-Traube rule [289, 299], according to which the surface activity of solutions raises by 3.2 – 3.5 times at transition to the higher homologue, is fulfilled. Fulfillment of this rule is the known illustration that the changes in the polymer strength are associated with reduction of the surface energy at the polymer-medium interface due to adsorption of medium molecules [287, 288].

PMMA durability was determined in acid solutions with concentrations, in which σ of this material is reduced in accordance with the Duklo-Traube rule. It has been found that for PMMA in acids (Figure 6.4) and in contact with the biomass (model experiments), $\lg\tau_d$-σ_d dependencies are of the analogous type. These dependencies are linear at definite stress levels (σ_d). This induces a suggestion that the change in PMMA durability in acid solutions (at stresses corresponded to linear parts of $\lg\tau_d$-σ_d dependencies) is of the adsorption origin [171, 172].

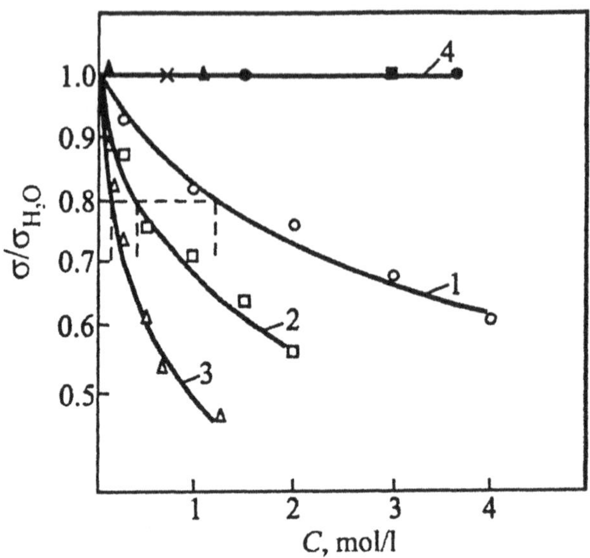

Figure 6.3. Dependence of PMMA induced elasticity limit ratio during elongation in aqueous solutions of acids and in water ($\sigma/\sigma(H_2O)$) on the acid concentration (C):
1 – acetic acid (O); 2 – propionic acid (□); 3 – butyric acid (Δ); 4 – tartaric acid (•), citric acid (■); oxalic acid (▲); succinic acid (×); and fumaric acid (♦); elongation rate equals 0.01 mm/min.

The analysis of experimental dependencies of PMMA induced elasticity limit and durability (linear parts of $\lg\tau_d$-σ_d curves) on concentrations of acid solutions has been carried out using the same approach, applied to the study of *Aspergillus niger*. If suppose that changes in mechanical properties of materials are proportional to the quantity of acid absorbed on its surface (a_c) and applying the Langmuir equation of monomolecular absorption, for every studied acid we obtain equation as follows:

$$\frac{C}{\Delta A/A_{H_2O}} = \frac{1}{\alpha_A a_p K} + \frac{C}{\alpha_A a_p} = p_A + n_A C, \qquad (6.3)$$

where C is the acid concentration in the solution; $\Delta A = A_{H_2O} - A_C$, A_{H_2O}, and A_C are the values of controlled index of PMMA mechanical properties in water and acid solution, respectively.

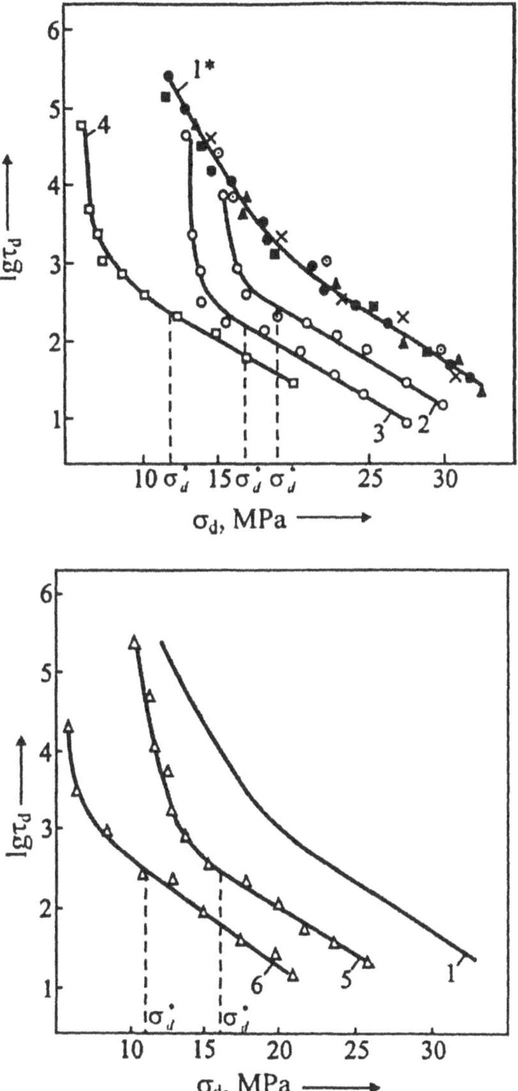

Figure 6.4. Dependence of PMMA durability (τ_d) on stress (σ_d) in: 1 - water (•) and 2-6 – aqueous solutions of acids:
1* - 0.5 mol/l succinic acid (■); 0.05 mol/l fumaric acid (⊙); 0.7 mol/l oxalic acid (▲); 3.0 mol/l citric acid (×); 3.7 mol/l tartaric acid (▲); 2 – 1.0 mol/l acetic acid; 3 – 2.0 mol/l acetic acid; 4 – 1.0 mol/l propionic acid; 5 – 0.2 mol/l; 5 – 0.2 mol/l butyric acid; 6 – 0.4 mol/l butyric acid.

If the adsorption hypothesis is true, mechanical parameters of PMMA in acid solutions should obey the Duklo-Traube rule. For all members of the homological sequence, n_A coefficient should be the same, because it is reversibly proportional to the capacity of dense adsorptive monolayer (a_p). In expression (6.3), the ratio of K values for consequent and previous homologues should be, approximately, 3.2 (at room temperature). The coefficient p_A should, respectively, be reduced by the same degree transiting to every consequent homologue. As a consequence for every used acid, it may be presented as follows:

$$p_A = \gamma_i \rho_A, \tag{6.4}$$

where γ_i is the coefficient considering surface activity of the acids-homologues, which is equal 1.0, 3.2, and 10.24 for dissolved butyric, propionic and acetic acids, respectively; i is the number of acid in the homological sequence; ρ_A is the coefficient constant for all homologues.

Transforming equation (6.3) with regard to expression (6.4), we obtain the generalized dependence of changes in polymer strength properties on the concentration of acid (the members of the homological sequence) solutions on it as follows:

$$\frac{C/\gamma_i}{\Delta A / A_{H_2O}} = \rho_A + \frac{C}{\gamma_i} n_A. \tag{6.5}$$

If changes in mechanical parameters of PMMA in media are determined by absorption of acid molecules to the polymer surface, τ_d-C and σ-C dependencies generalized for all applied acids should obey equation (6.5), and their graphs presented in coordinates of equation (6.5) should be linear.

Treatment of the experimental results indicates observance of this condition (Figure 6.5). For intervals of stresses exceeding some σ_d^* values, shown in Table 6.2 (linear parts of the durability curves in Figure 6.4), the correlation index of experimental τ_d-σ_d dependencies to equation (6.5) equals 0.95. Correspondence of experimental σ-C dependencies to the mentioned equation is estimated as 0.92, but just for a concentration range, definite for every acid shown in Table 6.2. At concentrations exceeding C^* value typical of every acid (Table 6.2, Figure 6.5), the correlation index is much lower.

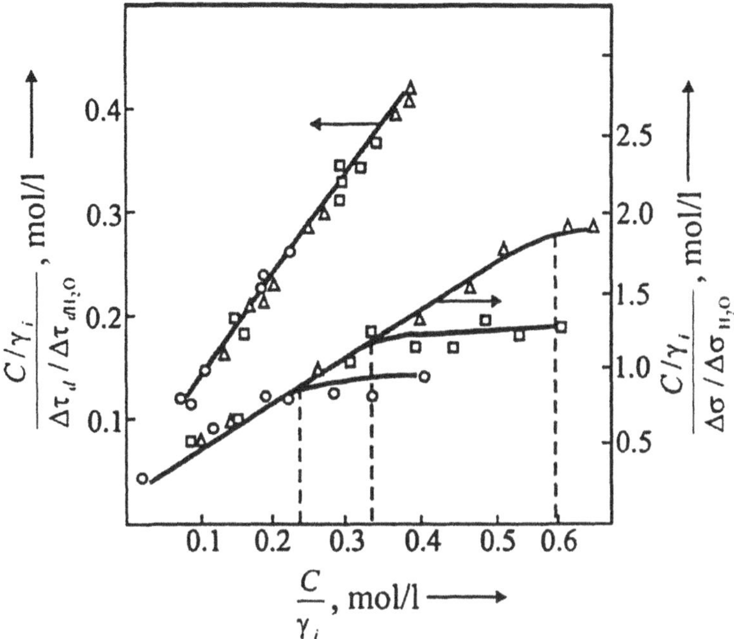

Figure 6.5. Dependence $\frac{C/\gamma_i}{\Delta A/A_{H_2O}}$ for solutions of acetic (○), propionic (□), and butyric (Δ) acids.

Hence, linearity of the graphic dependencies shown in Figure 6.5 is disturbed that testifies about a change in the mechanism of acid effect on PMMA strength. Taking into consideration that the initial elasticity modules of the polymer deformation in water and acids are equal, one may suppose stipulation of the change in PMMA strength in solutions with $C > C^*$ due to its local plasticization (mentioned in the consideration of the biodestructor effect on PMMA) in the area before the propagating crack tip.

Hence, experimental data on C and σ_d ranges, mentioned in Table 6.2, correlate with the Duklo-Traube rule. Table 6.2 shows parameters of equations (6.3) and (6.5) for changes of the PMMA strength in studied acids. Clearly K values calculated independently from the results of PMMA induced elasticity and durability limits determination in acidic solutions are close. As mentioned above, the independence of this coefficient on the mechanical test mode correlates with its physical meaning reflecting, in accordance with equation (6.3), properties of the surface-active substances.

Table 6.2

Values of equation (6.3) and (6.5) parameters for changing PMMA properties in organic acids

Coefficient *K*		
Acetic acid	**Propionic acid**	**Butyric acid**
Tests in constant rate elongation mode $\rho_\sigma = 0.22$; $n_\sigma = 3.4$		
0.2 mol/l $\le C \le C^* = 2.2$ mol/l 1.6	0.1 mol/l $\le C \le C^* = 1.2$ mol/l 5.1	0.1 mol/l $\le C \le C^* = 0.6$ mol/l 16.3
Tests in durability mode $\rho_{\tau_d} = 0.052$; $n_{\tau_d} = 0.85$		
$C = 1.0$ mol/l; 30.0 MPa $\ge \sigma_d \ge \sigma_d^\bullet = 19.0$ MPa	$C = 0.5$ mol/l; 27.0 MPa $\ge \sigma_d \ge \sigma_d^\bullet = 16.0$ MPa	$C = 0.2$ mol/l; 27.0 MPa $\ge \sigma_d \ge \sigma_d^\bullet = 16.0$ MPa
$C = 2.0$ mol/l; 27.5 MPa $\ge \sigma_d \ge \sigma_d^\bullet = 17.5$ MPa	$C = 1.0$ mol/l; 20.0 MPa $\ge \sigma_d \ge \sigma_d^\bullet = 12.5$ MPa	$C = 0.4$ mol/l; 21.0 MPa $\ge \sigma_d \ge \sigma_d^\bullet = 12.0$ MPa
1.5	4.8	15.5

Generally, test results for the model media show that in the concentration ranges definite for every acid and stresses applied to the material, changes in induced elasticity and durability of PMMA are caused by reasons of adsorption.

Comparative analysis of these results and changes in properties, obtained during tests, for materials contacting with microorganisms allows a conclusion that during formation (biomass growth) of a colony, its effect on strength and dielectric indices is stipulated by absorption of microorganism metabolites on the material. Adsorption reduces the energy of new surface formation at deformation and increases electric surface conductivity of polymers.

6.1.2. Sorption of microorganism metabolites in the polymer volume

Permeation of an aggressive medium into the material induces its plasticization. It is usually resulted in increasing volumetric current conductivity and changing strength properties of the polymer [171, 172, 302 – 307]. The plasticizer medium is capable of both increasing and reducing strength. Strength increase is usually corresponded to regulating the structure of polymer plasticized by the medium and simplifying over-stress relaxation in its defect areas. Strength reducing effect occurs due to weakening intermolecular interactions and (or) occurrence of internal (local) over-stresses and defects (cracks) due to irregular swelling.

It has been found in experiments reproducing the type of the real microbiological damage that the effect of reversible change in properties of some polymeric materials is displayed both in the presence and after removing of biomass from the samples. This fact allowed the supposition binding the mentioned effect first to the plasticizing action of metabolites diffusing into the material volume. Such reversible changes in strength of PE, PVC, PMMA, and CTA are observed, and dielectric properties – for varnished fabric and glass-fiber laminate.

Obviously, under the test mode used, metabolites in the sample volume are transferred almost uniformly along the surface contacting with the biomass. At the same time (see Chapter 1), the mode of metabolite transfer may be different in case of simultaneous effect of microorganisms and stretching stresses on the polymeric material. In this case, stress-activated diffusion of the medium (metabolites) may be localized and induce plasticization ("local plasticization") in the area before the tip of propagating crack (stress concentrator), thus changing the propagation rate of it [308 – 313].

Let us discuss the features of changes in properties induced by a biodestructor, associated with the above-mentioned types of the sorption-desorption processes proceeding in polymers.

In relation to the authors' point of view [283, 312 – 315], for the majority of polymer-medium couples, local plasticization and adsorptive reduction of the surface energy of the material are connected processes. Each of them is most clearly displayed under typical conditions of deformation, stretching stresses, elongation rate, etc. The results of analysis discussed in Section 6.1.1 allow specification of the test modes, at which local plasticization of PMMA by *Aspergillus niger* metabolites is the most probable reason of changes in its strength. Remind that when the stage of the exponential growth of this fungus on the material is over and biomass exceeds the m_A* level (Table 6.1), σ-m and τ_d-m dependencies presented in the Langmuir equation coordinates deviate from linearity (at $\sigma_d = 25$ MPa, Figure 6.2). Moreover, tests on durability (Figure 6.6) have indicated that if PMMA contacts with the biomass below the m_A* level, but is affected by the medium and comparatively low constant mechanical stresses, σ_d ($\sigma_d < \sigma_d$*), then the lgτ_d-σ_d dependencies do not obey the Zhurkov equation. This is confirmed by the presence of curved areas on the appropriated graphs. Table 6.3 shows values of σ_d*.

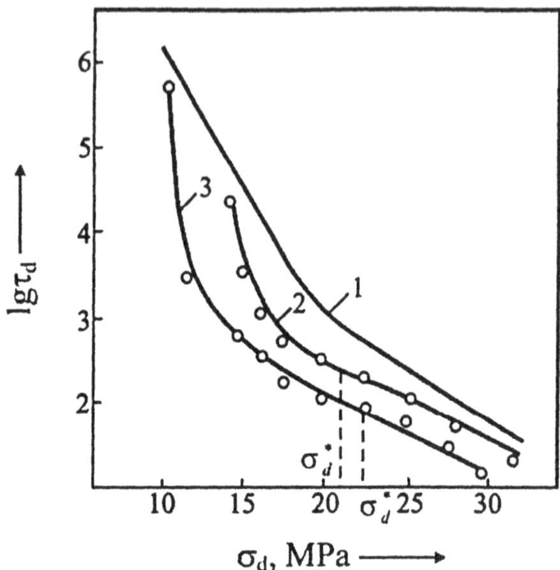

Figure 6.6. PMMA durability (τ_d) curves under *Aspergillus niger* effect: 1 – in humid atmosphere (control sample); 2 – after 216 hours of influence; 3 – after 336 hours of fungus cultivation on the polymer.

Table 6.3

Parameters of PMMA durability equation (6.6) in contact with *Aspergillus niger*

Duration of polymer contact with *Aspergillus niger* (*t*) and biomass (*m*)	Stress interval (σ_{dp} - σ_d*), MPa	σ_{dp} (theoretical), MPa	K_d, MPa·s$^{-1/2}$
$t = 216$ h; $m = 0.17 \cdot 10^{-3}$ mg/cm^2	14.8 – 21.5	13.2	127.7
$t = 336$ h; $m = 0.85 \cdot 10^{-3}$ mg/cm^2	10.8 – 23.5	10.0	130.1

The mentioned features of σ-m, τ_d-m, and lgτ_d-σ_d dependencies state that at the definite biomass and elongating stress values the changes in PMMA strength properties are not resulted by the metabolite adsorption effect only.

At the same time, these data may be explained using notions about local plasticization of polymers. At biomasses below m_A* and respective low concentrations of metabolites, as well as at relatively high stresses ($\sigma_d > \sigma_d$*) applied to the sample and, consequently, high rates of the crack propagation, the medium cannot diffuse (or diffuse in low quantities) in the material volume before the crack tip. As shown in Section 6.1.1, in this case, the thermal fluctuation mechanism of degradation is realized and PMMA strength reduction due to contact with *Aspergillus niger* are of the adsorption type (Figures 6.2 and 6.6, linear parts of the graphs).

On the contrary, relatively high biomasses (high metabolite concentrations) and (or) low stresses (low crack propagation rates) induce metabolite penetration into the material and its plasticization in a thin layer around the propagating crack tip. The effect of mechanical stresses on locally plasticized polymer orients molecules in this thin layer, which promotes its strengthening and deviation of $\frac{m}{\Delta\tau_d/\tau_{d0}}$-$m$, $\frac{m}{\Delta\sigma/\sigma_0}$-$m$, and $\lg\tau_d$-σ_d dependencies from linearity to the side of higher σ and τ_g values.

Suggested mechanism of PMMA degradation due to local plasticization is quite similar to the notions about steel hydrogen-stress cracking caused by stretching stresses [316, 317]. In accordance with these ideas, cracks develop in steels in the presence of hydrogen only, when its concentration in the metal reaches a maximum. Crack development with time will follow motion of the hydrogen diffusion front of the mentioned concentration. The typical feature of this mechanism is existence of the so-called "safe stress", i.e. the stress inducing no hydrogen-stress cracking of steel.

The quantitative description of experimental τ_d-σ_d dependencies, obtained by the authors (Figure 6.6 at $\sigma_d < \sigma_d^*$), were carried out using an analytical model analogous to the equation of hydrogen-stress cracking of metals [316]:

$$\left(\sigma_d - \sigma_{dp}\right)\sqrt{\tau_d} = K_d, \tag{6.6}$$

where σ_{dp} is the safe stress; K_d is the constant.

It has been found that the experimental data (Figure 6.6) on stress and biomass intervals mentioned in Table 6.3 are successfully approximated by equation (6.6). Hence, the correlation index equals 0.95.

According to the ideas on the hydrogen-stress cracking of steels, K_d constant characterizes acceleration of the crack spreading ($1/\tau_d$) with the increase of elongating stresses (σ_d) [316]. The data shown in Table 6.3 display independence of this constant on the biomass and, consequently, on duration of microscopic fungus growth on the polymer, which precedes the durability tests. This means that combinations of stretching stresses and biomasses (metabolite concentrations) affecting the material, at which durability is changed due to local plasticization, give K_d value constant for every material-microorganism couple. As shown below, this conclusion is confirmed by the test results on PMMA durability in model media – solutions of organic acids.

Dependence (6.6) suggests the presence of the so-called safe stress, σ_{dp}. As the biomass contacting with the polymer increases (as metabolite concentration increases), the safe stress is reduced (Table 6.3). In analogue to hydrogen-stress cracking of steel, one may assume that in the studied system, σ_{dp} means stress, at which metabolites of the current concentration do not practically affect the PMMA durability. Actually, safe stresses calculated by equation (6.6) are close to the experimental ones, at which time before sample degradation caused by the contact with *Aspergillus niger* and in humid atmosphere (control samples) are almost the same (Figure 6.6, Table 6.3). Hence, the calculation results confirm the supposition that the mechanism of PMMA degradation at contact with microorganisms is similar to the stress cracking of steel.

Equation (6.6) reflects regularities of PMMA durability changing at a definite moment of contact with the biomass at the stage of its irregularly accelerated and exponential growth at different loads applied to the polymer. Table 6.3 shows that every moment of time (biomass) possesses its own σ_{dp} value. One may suggest that the safe stress (σ_{dp}) dependence on time of microscopic fungus growth on the polymer (t) preceding durability tests will obey equation (5.4) of the biomass growth. Applying the growth equation to calculation of σ_{dp}, we obtain:

$$\sigma_{dp} = \sigma_{dp}{}^{*}\{1 + a\cdot\exp[-b(t - L)]\}, \tag{6.7}$$

where $\sigma_{dp}{}^{*}$ is the safe stress at a biomass value close to $m_{\tau}{}^{*}$.

Then according to expression (6.6), the durability (τ_d) kinetic equation under conditions of microscopic fungus biomass growth on the polymer (if $\sigma_d{}^{*} > \sigma_d > \sigma_{dp}$) will be as follows:

$$\sigma_d - \sigma_{dp}{}^{*}\{1 + a\cdot\exp[-b(t - L)]\} = \frac{K_d}{\sqrt{\tau_d}}. \tag{6.8}$$

Calculation results obtained from this equation coincide well with the appropriate experimental data (Figure 6.6).

Equation (6.8) parameters represent indices characterizing changes in the polymer properties induced by microorganisms, if the determining role is played by the local plasticization of polymer by metabolites. K_d constant and $\sigma_{dp}{}^{*}$ index characterize specificity of the material degradation at the simultaneous effect of stretching stresses and biomass. These indices reflect both material resistance to degradation and ability of particular species of microorganism to induce this type of degradation. In this case, biomass growth indices, a and b, characterize kinetics of changes in properties and its connection to the features of microorganism growth on the material.

The considered hypothesis of the process origin defining strength reduction of PMMA contacting with the microorganism has been proved by tests in model media: water solutions of monobasic organic acids. Figures 6.3 – 6.5 show the main experimental data, obtained in these tests (Section 6.1.1).

Comparison of these data with the results of experiments with microorganisms (Figures 6.2 and 6.6) indicates a series of general regularities. For example, starting from particular concentrations of acids, C^*, as well as at biomasses exceeding the m_p^* level and at stretching stresses, σ_d^*, applied to the sample, changes in PMMA mechanical properties obtained, presented in coordinates of Langmuir and Zhurkov equations, respectively, deviate from linearity. Thus in these cases, changes in strength properties cannot yet be explained by adsorptive regulations only. This stipulates competence of comparing mechanisms of acid solutions' and microorganisms' impact on the polymer within the range of the mentioned concentrations, biomasses and stretching stresses.

At active elongation of PMMA in acids, experimentally obtained changes of induced elasticity stop obeying the Langmuir equation (6.3), if solutions with concentrations above C^* are used (Table 6.2, Figure 6.5). It has been indicated by special investigations that the acid concentration in solutions is also sufficient for the type of their sorption by PMMA. A graph of dependence of the solution volume (θ^K) absorbed by the polymer on the acid concentration is shaped as a curve with a minimum (Figure 6.7)[7].

At low acid concentration, PMMA swelling is also low (below 1%). As shown in Section 6.1.1, in the same concentration range typical of each acid, changes in polymer σ in solutions are of the adsorption origin (Table 6.2). Starting from a concentration (C_θ^*, Figure 6.7) typical of each acid, which corresponds to the minimum on the $\theta^K - C$ curve, the quantity of solution absorbed by PMMA increases rapidly and may reach 5 – 10%. C_θ^* concentration is reduced in the sequence as follows: acetic acid (2.2 mol/l), propionic acid (1.0 mol/l), and butyric acid (0.5 mol/l). Obtained C_θ^* values are close to acid concentrations in solutions, at which the $C/(\Delta\sigma/\sigma_{H_2O}) - C$ dependence deviates from linearity (Figure 6.5, Table 6.2). Since the quantity of solution absorbed by the polymer is sharply increased in this concentration range (at $C > C_\theta^*$), it is quite probable that the main role in changing PMMA

[7] The quantity of solution absorbed by the polymer (θ^K) was determined from experimental sorption curves, obtained at the maximal time of sorption, the same for all acids (240 hours).

properties during deformation is played by the stress-activated medium diffusion into the material.

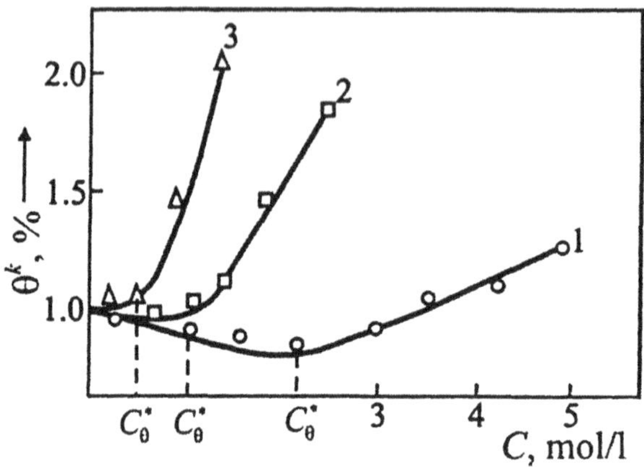

Figure 6.7. Dependence of the solution quantity (θ^K) absorbed by PMMA on concentration (C) of: 1 – acetic acid; 2 – propionic acid; 3 – butyric acid.

Table 6.4

Equation (6.6) parameters for PMMA durability in solutions of organic acids

Acid	Concentration in solution (C), mol/l	Stress range (σ_{dp} - σ_d*), MPa	Calculated σ_{dp}, MPa	K_d, MPa·s$^{-1/2}$
Acetic	1.0	14.0 – 19.0	13.8	66.7
	2.0	11.3 – 17.5	10.2	66.7
Propionic	0.5	9.0 – 16.0	8.3	125.8
	1.0	< 12.5	4.3	130.3
Butyric	0.2	10.0 – 16.0	9.7	160.6
	0.4	<12.0	3.1	153.7

Tests on PMMA durability in solutions with the acid concentration below C^* and stresses below σ_d^* (values of which are shown in Table 6.2) indicate $\lg\tau_d$ - σ_d graph deviation from linearity (Figure 6.4). In this range of stresses, $\tau_d = f(\sigma_d)$ and $\tau_d = f(C)$ dependencies cannot be described by Zhurkov's and Langmuir's equations and adsorption does not dominate in variations of the polymer durability anymore.

In this case, it is observed that the experimental data correspond to the analytical model (6.6), based on the previous suggestion about the main role of local plasticization in the zone before the propagating crack tip in PMMA degradation.

Table 6.4 shows parameters of equation (6.6), calculated for appropriate areas of the polymer durability curves in acid solutions. Parameter K_d is individual for each acid. As mentioned above, it characterizes crack spreading acceleration with stretching stresses. K_d is independent on the solution concentration, but increases at transition from acetic to butyric acid. Such regularity correlates well with the ideas about the mechanism of medium diffusion into the stressed material.

For example, it is indicated in ref. [311] that penetration of nitrogen molecules from the gas phase into the polymer before the crack tip requires a wave-like (oscillating) motion of a definite part of the polymeric chain. As the size of diffusing molecule increases, the greater part of the macromolecule is involved into the transport process [314].

Affected by stretching stresses, macromolecules receive additional energy and, consequently, they are involved into the transport process more easily, which accelerates medium diffusion into the material [310]. At the same applied stresses, the diffusion rate of acetic acid molecules, which are smaller than the molecules of propionic and butyric acids, will be higher. Assuming that the concentration, at which the crack propagates, is independent of the acid origin, the time for reaching this concentration in the zone at the crack tip will be shorter for acetic acid, and the crack propagation rate will be higher. As a consequence, acceleration of the crack propagation with σ_d increase, the same for all acids, will be higher for acetic acid, compared with propionic and butyric ones.

Values of the safe stress (σ_{dp}) calculated from equation (6.6) (Table 6.4) are close to the experimental stresses, at which acid in particular concentration causes no effect (compared with water) on the PMMA durability (Figure 6.4). Hence, obtained values of σ_{dp} and K_d correspond to the physical meaning of these parameters in the analytical model (6.6).

Generally, experimental results of the model media induce a conclusion that the main effect on PMMA strength in the range of comparatively high acid concentrations and middle and low stresses is caused by local plasticization of the material due to acid molecule diffusion, activated by the stress, into the zone of the polymeric body before the crack tip.

Analysis of experiments with microorganisms and model media gives basis to decide the fact proved that at simultaneous effect of comparatively low mechanical stresses and biomass after the end of exponential growth, changes in the material strength are controlled mostly by the stress-activated diffusion of metabolites into the zone before the propagating crack tip (local plasticization of the polymer).

If external mechanical stresses are absent, the type of the medium absorption by the material is changed. As mentioned above, in this case, the diffusing agent penetrates into the polymer from the total sample surface contacting with the medium. Then distributing in the material volume, the medium plasticizes it that induces changes in the operation properties.

Modeling of the real microbiological damage has indicated that such "uniform" diffusion of microorganism metabolites may stipulate reversible changes in strength and dielectric properties of polymers, observed at determination of appropriate indices after removing biomass from the surface of tested samples (Chapter 3). This mode of measuring electrical insulation resistance (R) was used by the authors of this monograph for obtaining dependence of this index of varnished fabric on time of *Aspergillus niger* cultivation on it (Figure 6.8). Kinetic curve of R changing can be conditionally divided into two areas. The first area (Figure 6.8, part I) is characterized by coincidence of experimental $R - t$ dependencies for the control samples and the ones infected by microbial cells. This means that the microorganism does not practically change the registered index. Thus it is changed by the effect of humid atmosphere, which concludes in water absorption by varnished fabric and consecutive change of electric conductivity [38, 318]. As shown in Figure 6.8, such reduction of R is observed during microorganism adaptation to the material and biomass growth up to ~ 0.1mg/cm^2. Then at a moment t_1, *Aspergillus niger* begins impacting on the insulation electric resistance (Figure 6.8, part II). It lasts continuously during growth of the microscopic fungus colony on the varnished fabric.

Experimental data may be explained as follows. It has been already mentioned that biomass growth is accompanied by the increase of metabolites' concentration on the material surface. As a consequence, diffusion and gradual accumulation in the varnished fabric volume proceeds that reduces its electric resistance. If the suggested mechanism is true, the change in R can be recorded (at the given sensitivity of the registration method) at accumulation of a definite quantity of metabolites in the varnished fabric only. The time period necessary for this depends both on the concentration of metabolites on the surface (the biomass quantity) and the intensity of their diffusion in the material volume. Apparently, the rate of R change is limited by diffusion, and the microorganism growth (metabolites' concentration increase in the solution on the sample surface) induces lower effect on it. This should explain the effect of somewhat "delay" in the change of electrical insulation resistance with regard to the biomass growth. The decrease of R is observed at $m \approx 0.1$ mg/cm^2.

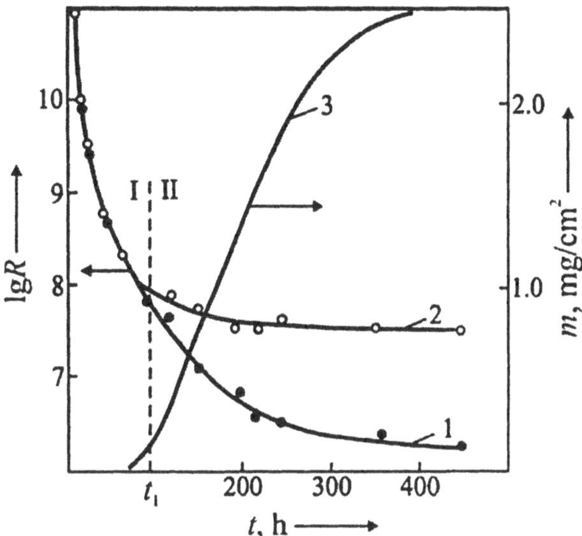

Figure 6.8. Dependence of electrical insulation resistance (*R*) of varnished fabric (1) and biomass quantity (*m*) of *Aspergillus niger* (3) on time of the fungus cultivation on the material; 2 – change of *R* in the control samples (temperature 29°C, relative humidity 98%).

It is common knowledge that the transfer of liquids in the material (if chemical interaction between them is absent) obeys the Fick equations. The known solutions of these differential equations [319, 320] were used by the authors for obtaining the analytical model of changes in properties, stipulated by the material plasticization due to biodestructor metabolite sorption in the volume.

To a first approximation, it has been assumed that the changes in electrical insulation resistance are directly proportional to the average concentration of the medium (microorganism metabolites) in the material ($\overline{C_m}$):

$$R_0 - R = \beta_R \cdot \overline{C_m}, \tag{6.9}$$

where R_0 and R are electrical insulation resistances before contacting the medium (microorganism metabolites) and at the moment t of the contact, respectively; β_R is a coefficient.

Using simplified solutions of the Fick equation [319, 320] describing changes in the substance concentration in the material volume with time and transiting from the metabolite concentration to R with the help of expression (6.9), we obtain:

- for short time of the medium impact (microorganism metabolites), i.e. initial stages of diffusion into the material ($F_0 < 0.2$)

$$\frac{R_0 - R}{R_0 - R_\infty} = 2\sqrt{\frac{F_0}{\pi}}, \tag{6.10}$$

where R_∞ is the minimal (equilibrium) electrical insulation resistance (R) reached under the medium (microorganism) impact; $F_0 = D_{eff}t/r^2$ is the Fourier criterion, where D_{eff} is the efficient coefficient of the medium (metabolites) diffusion into the material and r is the half thickness of the sample;
- for long times of the medium (metabolites) impact – at the final stage of diffusion ($F_0 > 0.2$)

$$\frac{R - R_\infty}{R_0 - R_\infty} = \frac{8}{\pi^2}\exp\left(-\frac{\pi}{4}F_0\right). \tag{6.11}$$

The results shown in Figure 6.9a testify about successful approximation of the experimental dependence $R = \varphi(t)$ at $t > t_1$ (Figure 6.8, curve I) by the models (6.10) and (6.11). The initial and final parts of this curve presented in coordinates of linearized expressions (6.10) and (6.11) are linear. Actually, calculations performed[8] have indicated that correspondence of the model (6.10) to experimental data at the initial part of the kinetic curve (at $F_0 < 0.2$) describing changes in R and that of the model (6.11) at the final part (at $F_0 > 0.2$) (Figure 6.8) is reliably estimated. For $F_0 < 0.2$, the correlation index equals 0.94, and for $F_0 > 0.2$ it equals 0.96.

Efficient diffusion coefficients calculated from equations (6.10) and (6.11) for linear parts of the graphs in Figure 6.9a are almost identical (Table 6.5). This is one more sign of the Fick law application to description of the studied process. It is known [320] that in cases when the diffusing agent is transferred with regard to the mentioned laws, the values of D_{eff} coefficient calculated from equations (6.10) and (6.11) should be equal.

[8] The value of electrical resistance (10^8 Ohms) at the moment t_1 is taken as the initial one (R_0) (see Figure 6.8).

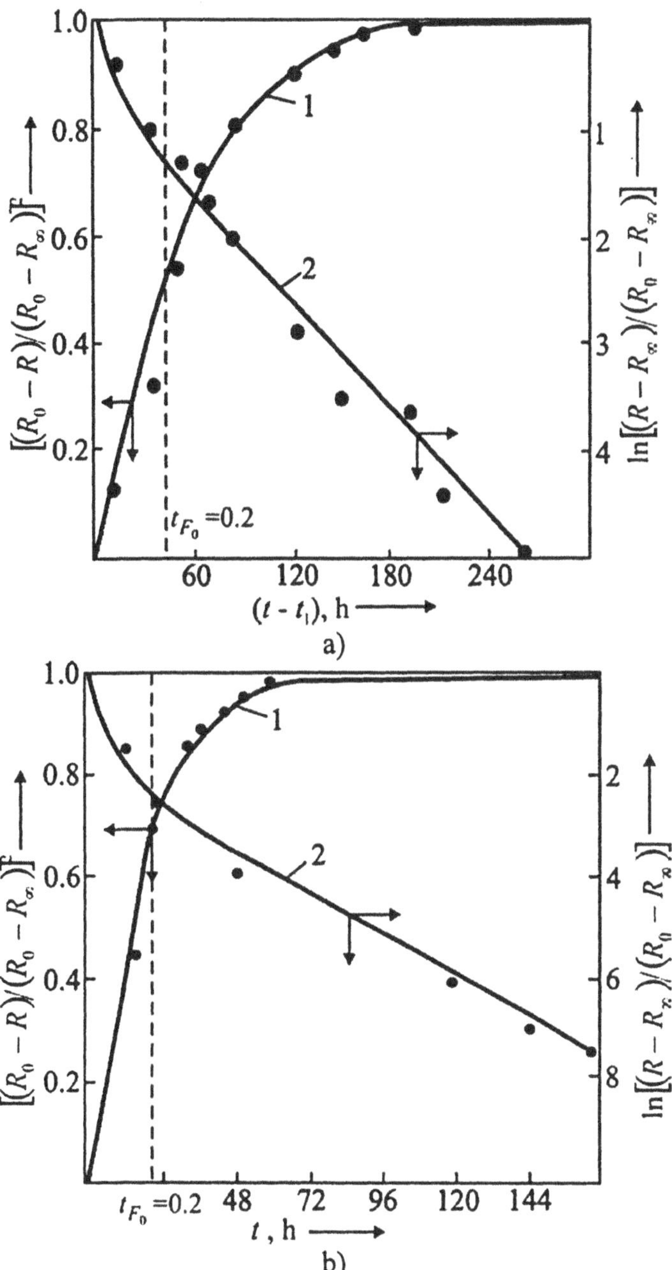

Figure 6.9. Kinetic dependencies of the insulation (varnished fabric) resistance (R) change in conditions of microscopic fungi cultivation (a) and control samples (b) in coordinates of linearized equations 1 - (6.10) and 2 – (6.11)

Table 6.5

D_{eff} coefficients from equations (6.10) and (6.11) for changes in electrical insulation resistance of varnished fabric

Medium	$D_{eff}\times 10^{12}$, m^2/s	
	$F_0 < 0.2$	$F_0 > 0.2$
Microscopic fungi	0.11	0.08
Humid atmosphere	0.27	0.24

The sorption-diffusion mechanism of metabolites' effect on electric resistance of varnished fabric is additionally proved by test results of samples treatment by non-inoculated fungus spores. The samples were exposed at 29°C and 98% humidity (control samples, Figure 6.8, curve 2). The data were treated by equations (6.10) and (6.11). It has been found that all the above-mentioned conditions testifying about validity of the Fick equations for describing changes in the controlled dielectric property are fulfilled. Thus in coordinates of linearized equations (6.10) and (6.11) appropriate graphs possess linear parts (Figure 6.9b), and diffusion coefficients calculated for every linear part are almost equal (Table 6.5). This means that the same conclusion [38, 318] about water diffusion into dielectric materials as the factor defining changes of electrical insulation resistance is induced by the analysis of calculation results.

Hence, the reversible change in varnished fabric properties at microscopic fungus growth obeys the Fick equation and is stipulated by the material plasticization, induced by sorption of fungal metabolites in the material volume. In this case, D_{eff} and R_∞ parameters of equations (6.10) and (6.11) can be used as the indices of changes in the properties, induced by the biodestructor. D_{eff} parameter characterizes the rate and R_∞ - maximally possible (equilibrium) change in the properties. The parameters are sensitive to both chemical composition of metabolites (the species of microorganism growing on the material) and their chemical and physical structure.

6.1.3. Plasticizer desorption from PVC-plasticate

Low-molecular component transfer from the polymeric material induces a change in its composition that may deteriorate strength, dielectric, optical and other important operation properties [171 – 175].

Modeling of the real microbiological damages has been displayed that long (ten-month) contact between PVC-plasticate and *Aspergillus niger* causes irreversible increase in the electrical insulation resistance (R was measured after biomass removing from the sample surface and conditioning under usual conditions). However, growth of electrical resistance (not so intensive) was also observed in the control samples exposed at 29°C and 98% humidity but not contacted with the microscopic fungus. This allowed association of microbiological degradation of dielectric properties with the processes analogous to those inducing these changes in PVC under thermal and humid conditions of aging.

Analysis of data from the literature [321, 322] has shown that such processes may be: chemical degradation of the polymer associated with dehydrochlorination and structuring (cross-linking) of macromolecules and plasticizer diffusive desorption to the environment. Hence, under conditions close to these used in the tests by the authors (low temperature, absence of UV-radiation), the plasticizer diffusive desorption dominates among processes of the material aging causing significant changes in operation properties, including rise of electrical insulation resistance. Moreover, the polymeric basis of PVC-plasticate is higher chemically and microbiologically resistant, but dialkylphthalate (DAP) used as a plasticizer is simply consumed by microorganisms [2, 76, 323, 324]. The data shown induce a suggestion that electrical insulation resistance rise observed during long impact of *Aspergillus niger* is caused by reduction of plasticizer concentration in the polymer.

Physicochemical tests aimed at clearing out this hypothesis were carried out. They concern studies of the composition and the structure of initial (affected by *Aspergillus niger*) and control PVC samples.

Possible chemical transformations of the insulation plasticate were determined by the IR-spectroscopy method in the spectrum zone typical of PVC-plasticate [325 – 328]. The results are shown in Table 6.6. Clearly absorption bands typical of polyvinylchloride peculiar to the C-Cl chemical bond are present in obtained spectra in 520 – 750 cm^{-1} frequency range. Thus intensity of these bands is approximately the same for all studied samples. All IR-spectra possess absorption bands in the ranges of 1650 and 2860 – 3000 cm^{-1}, typical of polyene sequences formed by PVC dehydrochlorination [325, 327]. Some quantity of polyene sequences does already exist in the primary insulating material [327]. One may also suggest that some insignificant quantity of them has been formed during thermooxidative degradation of the insulation (meaning 10-month exposure in the humid atmosphere at 29°C). At the same time, samples impacted by *Aspergillus niger* and the control ones display almost the same absorption intensity in discussed zones of the IR-spectrum.

Hence, the analysis of IR-spectrophotometry displays full identity of spectra from initial, control and fungus-infected samples of PVC-plasticate. All studied samples possess the same functional groups. As a consequence, dehydrochlorination of the polymer caused by microorganisms is low-probable.

Table 6.6

Typical absorption bands in IR-spectra of PVC-plasticate

Absorption band, cm^{-1}			Functional groups
Initial sample	**Control sample**	**Sample after microorganism impact**	
520 – 750	520 – 750	520 – 750	C-Cl
1650	1650	1650	CH_3-O-R; >C=O; >C=C<
2860 – 3000 (threshold)	2870 – 3000 (threshold)	2870 – 3000 (threshold)	RCH_2-(CH=CH_2); R-(C≡CH)

It is common knowledge that degradation of PVC-plasticate is accompanied by changes in its molecular-mass distribution (MMD) [325, 327]. Thus to confirm the absence of chemical transformations in the insulation during the contact with the biomass, MMD of the insulation was determined. The gel-penetrating chromatography method was used [327, 329].

It has been found that all the samples (initial, control and affected by microscopic fungi) display the bimodal distribution with the peaks, maximums of which correspond to confinement times equal 4.72 ± 0.03 and 7.96 ± 0.02 min. Appropriate values of the mid-viscous molecular mass are 4.8×10^4 (component A, the high-molecular one) and 3.6×10^4 (component B, the low-molecular one). The data shown in Table 6.7 indicate that the relation between low-molecular and high-molecular components is almost the same for all samples. As a consequence, the impact of *Aspergillus niger*, as well as thermal and humidity factors, induce no change in MMD of PVC-plasticate.

Table 6.7

Ratio of low- and high-molecular components in PVC-plasticate MMD

MMD component	Fractures of MMD components, %		
	Initial sample	**Control sample**	**Sample impacted by microorganism**
A	70.8	70.5	70.4
B	29.2	29.5	29.6

Hence, IR-spectrophotometry and gel-chromatography results show that chemical degradation of PVC-plasticate under the impact of microscopic fungi is low probable, and chemical structure of the material remains practically unchanged.

Possible plasticizer desorption process was experimentally determined by measuring dialkylphthalate (DAP) concentration in PVC-plasticate before and after tests. DAP concentration (C^{DAP}) in samples at a time t was determined by the UV-spectrophotometry method by intensity of the absorption band at $\lambda_{max} = 230$ nm, typical of this substance [328, 329].

It has been found that dialkylphthalate concentration in PVC-plasticate (C_0^{DAP}) is 27.2%. Ten-month impact of the temperature-humidity complex of factors ($T = 29°C$ and $\varphi = 98\%$, control samples) reduces DAP concentration in the plasticate to 25.8% (i.e. DAP concentration in the material is about 95% from the initial level). Samples impacted by the microorganism contain 20.4% of dialkylphthalate only, which gives 75% of the plasticizer in the initial PVC-plasticate.

As a consequence, dialkylphthalate concentration in the insulation is reduced under considered test conditions. Hence, compared with temperature and humidity factors (control samples), *Aspergillus niger* induces higher effect on DAP concentration.

Figure 6.10 shows kinetic curves of DAP relative concentration in PVC-plasticate (C^{DAP}/C_0^{DAP}) and electrical insulation resistance of the samples (R/R_0, where R_0 is the electrical insulation resistance of the initial PVC-plasticate) contacting with the biomass and impacted by temperature and humidity only. Clearly C^{DAP}-C_0^{DAP} and R/R_0 curves, obtained for the material impacted by *Aspergillus niger*, are analogous to those for the control samples. Reduction of the plasticizer concentration is accompanied by an increase of the insulation resistance.

If diffusive desorption dominates in changing electrical insulation resistance during its contact with the biomass, there should be a general dependence binding R/R_0 and C^{DAP}/C_0^{DAP} values obtained for the samples both impacted by microorganisms and aged by temperature and humidity. Calculations performed display validity of this condition. Figure 6.11 shows experimental dependencies, $R/R_0 = f(C^{DAP}/C_0^{DAP})$, obtained in tests on the samples both infected by microbial cells and control ones, as an uniform straight line. As a consequence, one may suggest that irreversible microbiological degradation of PVC-plasticate dielectric properties is stipulated by diffusive desorption of the plasticizer from the material.

Quantitative description of the studied process has been carried out using the known analytical models of the diffusive desorption of low-molecular components from polymeric materials [330 – 332]. Such models and their parameters allow estimation of changes in the concentration of desorbed component in the material with time. Using the correlation graph shown in Figure 6.11, they can be easily applied to determination of the changes of one property or another, if they are induced by the component migration in the material.

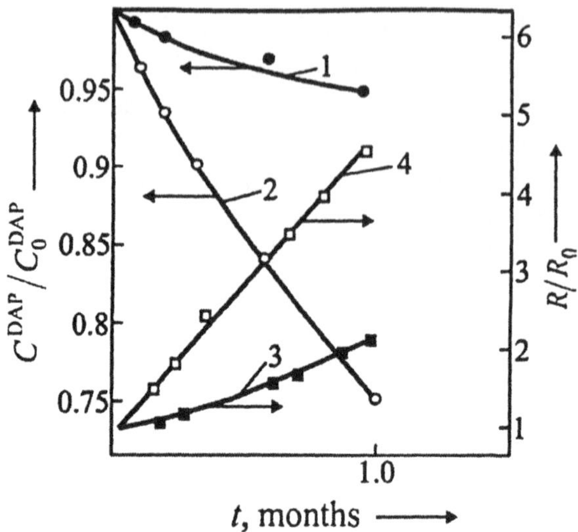

Figure 6.10. Kinetic dependencies of relative change in dialkylphthalate concentration (C^{DAP}/C_0^{DAP}) – 1, 2 and electrical insulation resistance of PVC-plasticate (R/R_0) – 3, 4 impacted by *Aspergillus niger* (○, □) and control samples (•, ■)

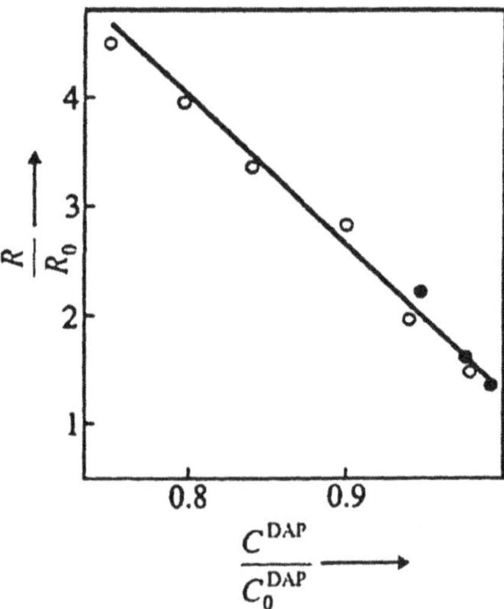

Figure 6.11. Dependence of the insulation resistance (R/R_0) on dialkylphthalate concentration in PVC-plasticate (C^{DAP}/C_0^{DAP}): (O) – impacted by *Aspergillus niger*; (•) – control samples.

As shown in refs. [330 – 333], regularities of variations of the plasticizer concentration in the plasticate are defined by the relation between rates of its diffusion to the material surface and desorption from it into the environment. For the majority of the material–migrating low-molecular component systems, one of the following two variants of the process proceeding is realized.

If the rate of desorption from the surface is lower than that of the low-molecular component in the material volume, then total rate of the process is limited by intensity of the plasticizer desorption to the environment (the so-called desorption zone of the process proceeding). For the desorption zone, kinetic equation describing changes of plasticizer concentration in the PVC-plasticate is as follows:

$$\frac{C^{\mathrm{DAP}}}{C_0^{\mathrm{DAP}}} = \exp\left(-\frac{W}{S \cdot r}t\right), \tag{6.12}$$

where W is the desorption rate of the component from the material surface; S is the solubility of the component in the material; r is the half thickness of the sample (film from the material).

In case, if the component is desorbed much faster than diffuses in the material, total rate of the process and, consequently, the quantity of migrating low-molecular component will be determined by diffusion (the diffusion zone). In this case, kinetics of the plasticizer concentration change in the material obeys the equation as follows:

$$\frac{C^{\mathrm{DAP}}}{C_0^{\mathrm{DAP}}} = \frac{8}{\pi}\exp\left(-\frac{\pi^2}{4}\frac{D}{r^2}t\right), \tag{6.13}$$

where D is the diffusion coefficient of the component in the material.

Equations (6.12) and (6.13) can be reduced to the equation as follows:

$$\frac{C^{\mathrm{DAP}}}{C_0^{\mathrm{DAP}}} = \exp\left(-k^{\mathrm{DAP}}t\right), \tag{6.14}$$

where k^{DAP} is the efficient rate constant of the low-molecular component transfer in the material.

Calculations have shown that experimental $C^{\mathrm{DAP}}/C_0^{\mathrm{DAP}} = f(t)$ dependencies (Figure 6.10) are approximated well by equation (6.14). Such formal-kinetic description enables describing the diffusion zone of dialkylphenyl desorption from the material and, consequently, to choose from equations (6.12) and (6.13) the one characterizing the features of studied microbiological degradation most well.

For this purpose, sorption-diffusion parameters D and W, obtained from the experimental data shown in Figure 6.10 and detected by other independent methods, have been compared.

Table 6.8 shows rate constants k^{DAP} of equation (6.14), calculated from the experimental data, and appropriate D and W parameters calculated from these values by equations (6.12) and (6.13).

Table 6.8

D and W parameters for the PVC-plasticate-dialkylphthalate system

Parameter	Sorption-diffusion parameters		
	Calculated from the experiment (Figure 6.10)		Calculated from known sorption-desorption characteristics
	Microorganism impact	Temperature impact (29°C)	Temperature impact (29°C)
Efficient rate constant (k_{eff}), c^{-1}	$1.1 \cdot 10^{-8}$	$1.6 \cdot 10^{-9}$	—
Diffusion coefficient (D), cm^2/s	$0.64 \cdot 10^{-10}$	$0.82 \cdot 10^{-11}$	$1.1 \cdot 10^{-10}$
Desorption rate (W), $g/cm^2 \cdot s$	$1.5 \cdot 10^{-10}$	$2.4 \cdot 10^{-11}$	$4.3 \cdot 10^{-11}$

Note: In calculations of the diffusion coefficient by equation (6.13), DAP solubility in PVC-plasticate was assumed equal 0.3 g/cm^2 [330].

Independently, these parameters have also been calculated from the data on physicochemical and sorption-diffusion characteristics of DAP and PVC-plasticate, present in the literature.

Dialkylphthalate desorption rate from the insulation surface was calculated by the Hertz equation as follows:

$$W = \frac{p_T}{\sqrt{2\pi M^{DAP} kT}}, \tag{6.15}$$

where p_T is the pressure of plasticizer vapor at temperature T; M^{DAP} is the DAP molecular mass (390 g-mole); k is the Boltzmann constant.

DAP vapor pressure at 29°C (p_{29}) was calculated with the help of interpolation dependencies obtained from the Clapeyron-Clausius equation:

$$\lg p_T = -A/T + B, \tag{6.16}$$

where *A* and *B* are constants.

To determine constants *A*, *B*, and p_{29}, a system of equations of the (6.16) type was solved at 293 and 332 K. Plasticizer pressures at these temperatures ($1.3\cdot10^{-4}$ dyn/cm^2 and $4.0\cdot10^{-3}$ dyn/cm^2, respectively) are shown in ref. [330]. The calculation has indicated that at 29°C, DAP vapor pressure equals $3.4\cdot10^{-3}$ dyn/cm^2. Table 6.8 shows also the desorption rate calculated by equation (6.15).

The coefficient of DAP diffusion from PVC-plasticate at 29°C was calculated using data from ref. [333], in which values of the coefficient *D* for DAP in the insulation plasticate are shown for different temperatures. Using the Arrhenius equation, the efficient activation energy of diffusion was calculated (58.4 kJ/mol), and then the *D* value at 29°C, shown in Table 6.8, was calculated.

Analysis of the data from Table 6.8 shows that for *Aspergillus niger* impacted PVC-plasticate, experimental (Figure 6.10) and calculated values of *D* coincide well. At the same time, desorption rates differ almost by an order of magnitude. So it may be concluded that under microbiological impact, the plasticizer transfer from PVC-plasticate obeys equation (6.13), and the total rate of the process is defined by DAP diffusion in the material volume.

Table 6.8 shows also parameters from equations (6.12) and (6.13), obtained from the experimental data (Figure 6.10) on the plasticizer transfer, affected by temperature and humidity only (T = 29°C, φ = 98% -control samples). These desorption rates and the calculated ones are almost equal. As a consequence, as insulation from PVC is affected by temperature (29°C) and humidity (98%), the plasticizer leaving rate is first determined by its desorption from the material surface. This output correlates well with the data from ref. [330] which show that at low-temperature aging, diffusion desorption proceeds in the desorption zone. Such correspondence of the experimental data and the ones from the literature proves adequacy of the methodological approach to identification of the plasticizer migration type from the insulation plasticate used.

Hence, the data obtained prove that under long-term effect of microorganisms on PVC-plasticate, the diffusion desorption dominates and leads to irreversible changes in the dielectric properties. Hence, under favorable conditions for the microorganism growth (T = 29°C, φ = 98%), the plasticizer yield is limited by its diffusion in PVC-plasticate volume. In this case, the diffusion coefficient of the low-molecular component in the material volume and the correlation coefficient (correlation dependence) binding the value of controlled properties with migrating component concentration in the material indicate microbiological degradation of the properties.

Note that microscopic fungi inhabiting PVC-insulation may possess lower ability to assimilate DAP, compared with the used *Aspergillus niger* strain, or external conditions will be less favorable for the microorganism growth. Thus the zone of diffusive desorption will be different, for example, desorption or mixed diffusion-desorption one. In this case, determining microbiological degradation indices induced by the low-molecular component transfer from the material, it is desirable to determine first the zone of the process proceeding for every particular microorganism-material couple.

6.1.4. Biomass contamination of fuels and lubricants

Growth of microorganisms in fuels and lubricants induces changes in the material quality, for example, concentration of mechanical admixtures. Accumulated biomass causes the highest effect on operational properties of oil-distillate fuels, lubricating oils and special fluids (Chapter 3).

Microscopic fungi and bacteria growth and, consequently, increase of mechanical admixture concentration in fuels and lubricants are discussed in detail in Chapter 5. Its kinetics is described by exponential equation (5.4).

Parameters of the equation can also be used for estimating changes in mechanical admixture concentration. They consider the process rate and possible effect of the microbiological factor on this parameter of fuels and lubricants.

6.2. CHANGES IN MATERIAL PROPERTIES INDUCED BY CHEMICAL PROCESSES

Reproduction of the character of the real microbiological damage of materials has induced a supposition that irreversible changes in properties of varnished fabrics, cotton threads, polysulfide hermetic sealer, as well as aluminum and steel alloys are caused by chemical (electrochemical) transformations proceeding in the material under the impact of microorganisms. Analysis of these results from positions of chemical aggressive media impact on the material allowed a conceivable identification of the processes' type inducing changes in operation properties. These processes are: chemical degradation of polymeric materials and electrochemical corrosion of metals caused by metabolites of microorganisms.

6.2.1. Chemical degradation of polymeric materials induced by metabolites of microscopic fungi

Chemical degradation induces changes in chemical structure and molecular mass of polymers, accompanied by irreversible changes in operational properties. It includes sorption of aggressive medium and future reactions of chemical unstable bonds' transformation in the material [2, 171, 174, 175].

In experiments on physical modeling of biodamages, irreversible changes of cotton threads, hermetic sealer U-30MES-5 and varnished fabric was observed after biomass removing from the samples and their conditioning. Such mode of measurements has maximally excluded the influence of metabolites' physical sorption by materials on strength and electrical resistance. This allows a supposition that in this case, chemical transformations in polymers are of the main role. Validity of the hypothesis about the leading role of chemical processes in the microbiological damage of cotton threads and varnished fabric is confirmed by numerous data present in the literature on the mechanisms of microorganisms' effect on cellulose and cellulose-containing materials [2, 274, 335]. Microscopic fungi and bacteria break glucosidic bonds in cellulose macromolecules, which is accompanied by the change in properties of materials containing cellulose in their composition. At the same time, it should be noted that ethyl cellulose varnish used in production of varnished fabric contains several low-molecular additives (plasticizer, stabilizer, and a dye). Consequently, there is a theoretical possibility of irreversible change in electrical resistance of the varnished fabric, which, besides chemical degradation, is also induced by transfer of the above-mentioned additives into the environment. However, reduction of the low-molecular concentration usually increases electrical resistance of dielectrics [318, 330, 331]. On the contrary, the authors of the present monograph have been found experimentally that the effect of *Aspergillus niger* on the varnished fabric is accompanied by reduction of the resistance.

A supposition about chemical origin of the processes defining changes in strength of polysulfide sealer required additional experimental proof. For this purpose, the change in the cross-linking degree (j) of the material was determined after impact of *Cladosporium resinae* microscopic fungus metabolites on it. It is known [174] that degradation of U-30MES-5 in chemically active media is accompanied by reduction of the cross-linking degree of molecules. The value j of the sealer before and after impact of microscopic fungus metabolites was determined with the help of the equilibrium swelling method (the sol gel analysis) [329]. It has been found that after 10 months of incubation in solution of microorganism metabolites (in the water phase of water-fuel-mineral solution during *Cladosporium resinae* development on it), the cross-linking degree is reduced by about 50% compared with its initial level (for the initial samples, $j = 0.47$). The value of this structural index obtained for control (incubated in the water phase of sterilized water-fuel-mineral medium) and initial samples are practically the same. Reduction of the cross-linking degree in samples contacting with *Cladosporium resinae* testifies about the break of chemical bonds in polysulfide sealer polymeric matrix.

Figure 6.12 shows kinetic curves of changes in strength and dielectric properties of studied polymeric materials impacted by microscopic fungi. Changes in indices were recorded after sample cleaning from biomass and conditioning.

Clearly irreversible changes in properties are initiated already during formation (development) of the colony (for cotton threads and sealer) and last during the whole contact of the polymer with the biomass (fungal metabolites). Hence, after 10 months of tests, the reduction (in relation to initial values) of cotton thread strength is up to 90%, ~60% for the sealer, and ~75% for the electrical insulation resistance.

Tests of control samples have shown that the effect of temperature and humidity only (T = 29°C, φ = 98%) also induces irreversible changes in properties of cotton threads and varnished fabric. However in this case, it is much lower intensive, than at polymer contact with a microscopic fungus. For the control samples, the change in varnished fabric resistance (R) is detected only 7 months after (already after 4 months of contact with *Aspergillus niger*), and reduction of σ of cotton threads – after 1.5 months (under microbiological effect – after 10 days) of tests. These indices (in relation to their initial values) are changed by 10-15%. No changes in strength of U-30MES-5 sealer control samples were observed. Hence, kinetic curves shown in Figure 6.12 reflect changes in polymer properties, mostly induced by the microbiological factor.

Of special attention is the fact that kinetic curves are of different shapes (Figure 6.12) that testifies about the difference in macrokinetic regularities of changes in properties of studied materials. Actually, these materials are of different chemical origin and physical structure. As a consequence, every of these materials are characterized by the self mechanism and rate of both sorption of chemically aggressive metabolites and transformation of unstable chemical bonds. Moreover, the origin and composition of metabolites inducing degradation are different for every species of microscopic fungus and type of the material. Thus it follows that every microorganism-material couple should be characterized by the unique kinetic regularities of degradation.

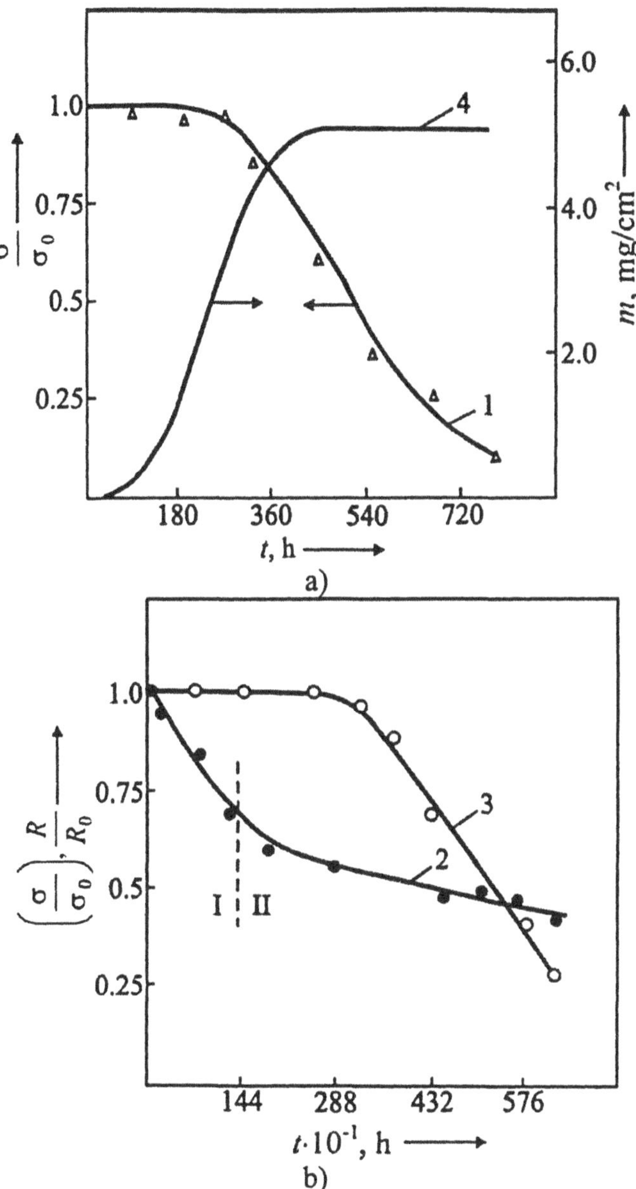

Figure 6.12. Kinetic curves of relative rupture stress reduction (σ/σ_0) for: a – cotton threads (1); b – U-30MES-5 sealer (2); electrical insulation resistance (R/R_0) of varnished fabric (3) impacted by *Aspergillus niger* and a – kinetic curve of the fungus biomass growth on cotton tarpaulin fabric (4).

Analytical models of irreversible changes in properties have been composed using an approach developed for studying the effect of chemically aggressive media on polymers [175].

According to ref. [175], the regularities of changes in the controlled parameter of polymer properties in chemically aggressive media are defined by the medium diffusion rate in the material volume and chemical transformations. Refs. [169, 336, 337] indicate quite wide areas for application of this approach to description of biologically active medium impact on materials.

In case, when chemical reaction proceeds much faster than the medium diffusion in the material (external diffusion-kinetic zone), degradation represents the diffusion-restricted process proceeding in a thin sub-surface layer of the polymer.

If the medium diffusion rate in much higher that the chemical reaction rate, then degradation proceeds in the whole volume of the material simultaneously after the end of its swelling in the medium (the internal kinetic zone).

For chemical degradation proceeding in the external diffusion-kinetic zone, changes in properties obey the equation as follows:

$$A = A_0(1 - k_{\text{eff}}t)^2, \tag{6.17}$$

where A and A_0 are values of controlled property index before (initial) and after contact of the material with the aggressive medium during time t, respectively; k_{eff} is the efficient degradation rate constant.

Changes in the material properties induced by degradation in the internal kinetic zone are described by the expression as follows:

$$A = A_0(1 - k^*_{\text{eff}}t), \tag{6.18}$$

where k^*_{eff} is the degradation rate constant.

Applicability of the models (6.17) and (6.18) to description of experimental data shown in Figure 6.12 has been approbated. The areas of the kinetic curves, in which changes of controlled characteristics are absent, were not considered.

It has been found that equation (6.18) is preferable as the analytical model of electrical insulation resistance reduction for varnished fabric. In this case, the correlation index equals 0.95, and for equation (6.17) it is 0.65. This means that varnished fabric degradation reducing its electrical resistance penetrates into the internal kinetic zone.

The change in strength of cotton threads is the most accurately described by equation (6.17) (the correlation index equals 0.97). Consequently, reduction of the rupture stress of cotton threads is first defined by chemical transformations in the material induced by its interaction with metabolites in the thin surface layer (the external diffusion-kinetic zone).

The kinetic dependence of changes in polysulfide sealer strength contacting with *Cladosporium resinae* metabolites is the most complicated.

Analysis of the dependence $\sigma/\sigma_0 = f(t)$ shown in Figure 6.12b (curve 2) displays two parts on it. The first part (curve 2, part I) represents the initial stage of the process. It is linear. As a consequence, one may suggest that at the initial stage of metabolites' impact, degradation of the sealer obeys equation (6.18), and the total rate of the process is defined by the chemical reaction rate. Degradation proceeds in the internal kinetic zone. As the contact time between the sealer and metabolites increases, the linearity of the curve $\sigma/\sigma_0 - t$ is disturbed (curve 2, part II). Obviously, in this case, the degradation rate will be defined by the diffusion rate of active components of the fungus metabolites in the material volume, and the process proceeds in the external diffusion-kinetic zone. Actually, calculations have indicated that the analytical models (6.17) and (6.18) approximate the experimental data on appropriate parts of the kinetic curve $\sigma/\sigma_0 - t$ quite well.

Table 6.9

Values of k_{eff} and k^*_{eff} constants from equations (6.17) and (6.18) for changes in properties of polymers impacted by microscopic fungi

Material-microorganism, controlled characteristic	Time period of contact with microorganism, hour	Type of kinetic equation	Efficient degradation rate constant, hour^{-1}
Cotton thread – *Aspergillus niger*, rupture stress	270 – 720	(6.17)	1.0×10^{-3}
Varnished fabric – *Aspergillus niger*, electrical insulation resistance	2,880 – 6,200	(6.18)	1.6×10^{-4}
Hermetic sealer – *Cladosporium resinae*, rupture stress	0 – 1,440 1,440 – 6,480	(6.18) (6.17)	8.3×10^{-5} 1.7×10^{-5}

Hence, equations (6.17) and (6.18) reflect adequately types of the processes proceeding during irreversible changes in properties of the studied materials contacting with the microorganisms. Calculated efficient degradation rate constants are shown in Table 6.9.

Generally, the results shown give grounds for a suggestion that an irreversible change in properties of polymers is mainly induced by chemical transformations in these materials caused by metabolites of microorganisms and obeys the regularities typical of the polymer degradation in liquid, chemically aggressive media.

Efficient degradation rate constants are recommended as the indices of such effect of a biofactor on the material. Moreover, the time period, during which no changes of the controlled index are observed, will also be unique for each microorganism-material couple. These indices as well as the information about the zone of the process proceeding fully characterize the considered stage of microbiological damage, when it is induced by chemical transformations in the material.

6.2.2. Electrochemical corrosion of St. 3 steel and D-16 aluminum alloy in contact with microorganisms

Experiments on modeling the real microbiological damage have displayed signs of corrosion, typical of metals, on samples of St. 3 steel and D-16 aluminum alloy contacting with microorganisms. A dark-brown film, separate ulcers and cavities were observed on the surface of the steel sample contacting with *Bacillus sp*. Light- and dark-gray spots and point damages were observed on D-16 alloy surface contacting with *Aspergillus niger*.

Analogous damage signs were observed on control samples. However, they appear and are developed with lower intensity. Damages of the control samples and used test conditions (φ = 98%, T = 29°C) reliably prove electrochemical origin of corrosion inducing the damages. Consequently, one may suggest that microorganisms also induce the electrochemical mechanism of corrosion in the humid medium.

This supposition has been proved by electrochemical studies of aluminum alloy and steel samples in water solutions of microorganisms' metabolites (cultural medium) and microbiological nutritious medium (control samples)[9].

Figure 6.13 shows the features of anodic behavior of steel in culture of *Bacillus sp*. bacteria. It is clearly observed that compared with the control test in the medium obtained after one day of bacteria development, the current of

---

[9] Polarization curves were obtained with the help of potentiometer P-5827. A glassy electrochemical cell with separate anode and cathode spaces was used. The sample tested was used as an electrode. The electrode for comparison was made from silver-chloride. Current was measured 5 minutes after setting the sequential potential. Cultural medium and microbiological nutritious medium were used as electrolytes. The cultural medium was obtained by cultivating microorganisms in liquid nutritious media with future biomass filtering from it. Chapter 2 shows the composition of the nutritious media.

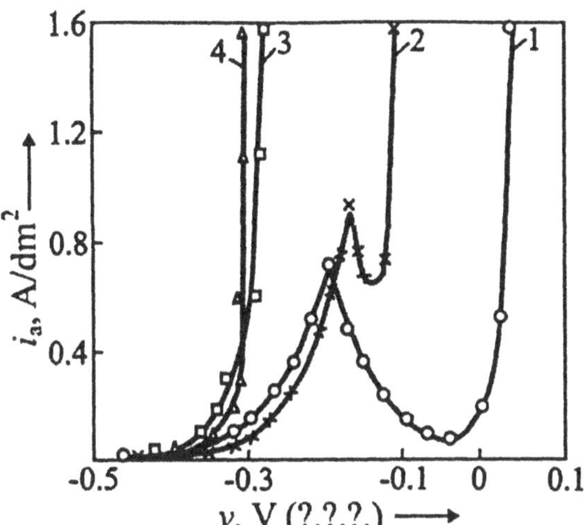

Figure 6.13. Anodic polarization curves of St. 3 steel in *Bacillus sp.* metabolites 1 – sterile nutritious medium (pH 7.0); 2 – after one day of cultivation (pH 6.6); 3 – after four days of cultivation (pH 4.2); 4 – after ten days of cultivation (pH 4.1); i_a is the current density; v is the potential.

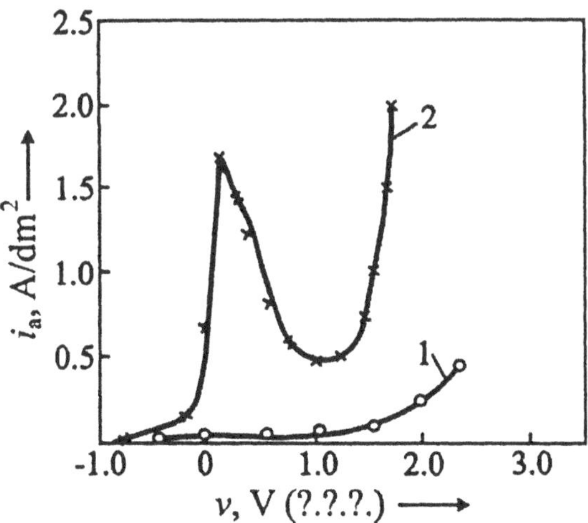

Figure **6.14.** Anodic polarization curves of aluminum alloy D-16 in *Aspergillus niger* metabolites

1 – sterile nutritious medium (pH 7.0); 2 – after four days of fungus cultivation in nutritious medium (pH 3.7).

active steel dissolution is noticeably increased (curve 2). The use of metabolites' solution, obtained after 4 and 10 days of *Bacillus sp.* cultivation displayed absence of the passive state of the metal on polarization curves 3 and 4 (Figure 6.13). In the nutritious medium (control), the anodic curve is shaped typical of steel corrosion in neutral media.

The change of the corrosion type is stipulated by accumulation of metabolites in the medium that increases the medium acidity. For example, after 10 days of bacteria metabolism, pH of the medium is reduced from 7.0 (nutritious medium) to 4.1 (culture). Hence, according to the Purbe diagram (potential – pH), steel corrosion in a single equilibrium zone with a passive state is shifted to another zone, where the passive state may not be reached.

Electrochemical test results on the aluminum alloy are shown in Figure 6.14. It has been found that the anodic curve obtained in the nutritious medium (control) with pH 7.0 is smooth, and the current density is increased at the potential shift to the anodic zone by more than 2.5 V from the given stationary potential. In accordance with the Purbe diagram for aluminum under conditions, in which the anodic curve is measured (at pH 7.0 of the medium and potential from -0.48 to +2.0 V), D-16 alloy exists in the passive state zone [338].

For D-16 alloy, the anodic curve in solution of *Aspergillus niger* metabolites is of the classic shape. Compared with the control potential, stationary one of the alloy is shifted by 0.3 V towards zero and equals 0.78 V. In the potential range from -12 to +0.17 V, a sharp increase of the anode current density is observed, and corrosion of the sample surface is inhibited and then over-inhibited at potentials above +0.17 V only. Such behavior of the metal in the culture is stipulated by the medium pH shift to higher acidity (pH 3.7) after 4 days of fungus cultivation that promotes destruction of the surface oxide film.

These results of polarization tests correlate well with the literary data on behavior of steel and aluminum alloys in electrolyte solutions [339], and occurring corrosion effects are associated with electrochemical processes.

Kinetic dependencies of corrosion depth in St. 3 steel and D-16 aluminum alloy caused by microorganisms are shown in Figure 6.15.

Analysis of the dependence obtained for steel 3 shows that the difference between corrosion depths of samples contacting with a biodestructor (*Bacillus* sp.) and control ones is practically constant (expect for initial linear parts of the curves). Corrosion induced by metabolites of microorganisms is mainly accelerated during initial 450 hours of tests. In both cases, future corrosion proceeds at the same rate that testifies about electrochemical type of the process for the samples both affected by microorganisms and control ones.

Approximately the same conclusion can be made about D-16 aluminum alloy samples. However, greater difference in the process rates in the initial period of time is observed in this case. In samples impacted by microorganisms

(*Aspergillus niger*), corrosion occurs after several initial hours of tests, whereas in the control samples it is observed after the incubation period (about 720 hours). The incubation period is explained by the presence of protective oxide film. After destruction of the oxide film on the aluminum alloy, microorganisms accelerate corrosion at the initial stage. Corrosion proceeds by the chemical mechanism in analogue to the temperature and humidity impact.

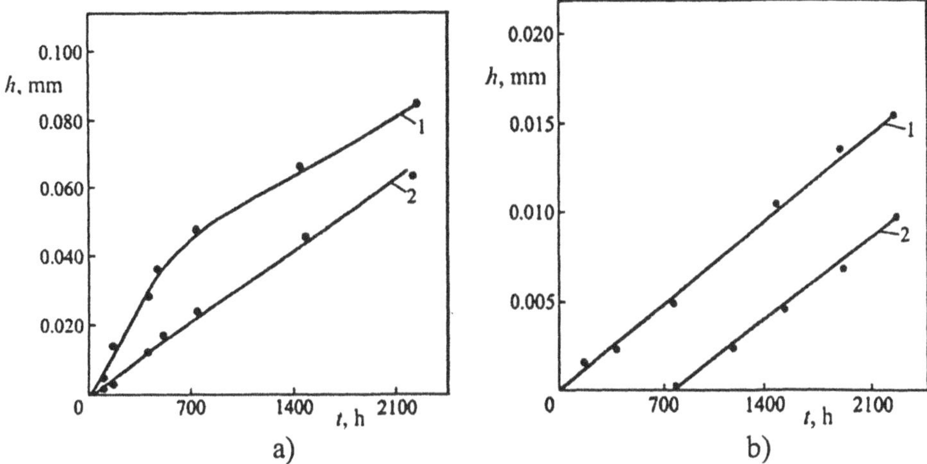

Figure 6.15. Dependence of the corrosion depth in: St. 3 steel (a) and D-16 aluminum alloy (b) samples on time of their contact with microorganisms.
1 – contact with microorganisms; 2 – control samples

Hence, investigations have indicated two (initial and sequential) periods in corrosion development of steel and aluminum samples. For the steel samples, microorganisms are active during initial 450 hours, when dependence of corrosion depth on the test time is linear. Further on, the dependence type is changed.

For aluminum alloys, the biodestructor effect eliminates the incubation period. Occurrence and increase of the corrosion damage is observed already during initial hours of tests. Hence, kinetic dependence of the damage depth is linear.

It has been found that linear parts of kinetic dependencies of steel and aluminum alloy damage obey a linear equation obtained in refs. [340 – 343], which describe electrochemical corrosion under atmospheric conditions at the initial period of tests (when the effect of corrosion products on its rate may be neglected):

$$h = k(t - t_1), \quad (6.19)$$

where h is the corrosion depth; k is the corrosion rate; t_1 is the incubation period of corrosion; t is the test time.

Table 6.10 shows calculated corrosion rates and incubation periods. Clearly they give quantitative estimation of the biofactor effect on corrosion.

The corrosion rate of St. 3 steel samples affected by microorganisms is 3-fold higher than that of the control ones. The corrosion rate of D-16 alloy is by an order of magnitude lower than for the steel samples. Hence, it is the same in the presence and in the absence of microorganisms. In this case, the action of biodestructor is characterized by the presence and duration of the incubation period.

Table 6.10

Equation (6.19) parameters for corrosion of St. 3 steel and D-16 aluminum alloy samples in the presence of microorganisms

Metal, microorganism	Time interval, hours	k, mm/hour	t_1, hour
St. 3 steel, *Bacillus sp.*	0 – 450	7.7×10^{-5}	0
St. 3 steel, control samples	0 – 1,440	2.5×10^{-5}	0
D-16 aluminum alloy, *Aspergillus niger*	0 – 2,160	7.0×10^{-6}	0
D-16 aluminum alloy, control samples	720 – 2,160	6.9×10^{-6}	715

Hence, it has been found that corrosion of metals induced by microorganisms and their metabolites obeys regularities typical of electrochemical processes. Hence, the rate of corrosion processes in the presence of microorganisms is much higher, especially during initial one or two months of tests. The corrosion rate and duration of the incubation period are the indices characterizing this stage of metal and alloy microbiological damaging.

The regularities, analytical models and suggested indices, studied in the current Section, reflect the origin and features of microbiological deterioration of the material properties and may give grounds to development of the forecasting methods. Appropriate kinetic models allow calculation of quantitative changes in the operation properties of materials in the presence of microorganisms under some (for example, usual) conditions of the environment. If the medium conditions (real operation conditions) are changed, then forecasting of the microbiological degradation impact should consider additional dependencies of changes in properties on different alternating external factors (temperature and humidity, for example). Chapter 7 discusses problems of mathematical modeling and forecasting of microbiological degradation in more detail.

The results obtained prove that changes in the material properties, induced by microorganisms, are mainly stipulated by the material interaction with aggressive media containing biodestructor metabolites. This gives grounds to a supposition that suggested indices of the considered stage of microbiological damage will be sensitive and can be used in estimation of safety measures and means efficiency, based on increasing material resistance to the aggressive media. Such methods and means are discussed in Chapter 1.

CONCLUSION

Analysis of the results discussed in the current Chapter shows that changes in the material properties induced by microorganisms obey regularities typical of interactions between liquid aggressive media and solid materials, aging of polymers and corrosion of metals.

Changes in properties are caused by the impact of microorganisms' metabolites on materials, induced by both physical and chemical processes. The physical processes are: sorption of metabolites by polymers, desorption of low-molecular components from polymers, as well as contamination of fuels and lubricants by biomass. Chemical processes are chemical degradation of polymers and electrochemical corrosion of metals. Physical processes may induce reversible and/or irreversible changes. Chemical ones induce irreversible changes only.

Origin of a process defining change of a property depends on the material type, species of microorganism and interaction conditions.

Analytical dependencies reflecting the mechanism and kinetics of biodestructor impact on properties of various materials and associating changes in properties with preceding stages of biodamaging (adhesion of microorganisms and their growth) are obtained. The possibility to use parameters of dependencies as indices of studied stage of microbiological damaging is proved. Suggested indices allow quantitative estimation and forecasting of the material resistance to biodestructor and efficiency of protection measures.

Obtaining of these indices with the help of techniques used in studies of aggressive medium impact on materials, corrosion of metals and aging of polymers and providing for determining kinetics of appropriate processes under the effect of different factors on the material defining type and intensity of the process under real operation conditions, are proved experimentally.

Chapter 7.

Protection of materials and technical facilities against microbiological damage

---

Reliable protection of technical facilities against unfavorable impact of a biofactor may be achieved by a complex of measures developed within the frames of the uniform system of activities, based on data on microbiological resistance of materials, articles, protection facilities and methods, and forecasting of its changes under operation conditions.

The ideas developed in the present monograph consider microbiological damage of materials as a multistage process. Parameters, indices and analytical models suggested for its study and description of quantitative regularities are realized in appropriate techniques and recommendations on carrying out investigations and forecasting microbiological resistance, development and application of materials and articles protection facilities and methods.

7.1. SYSTEM APPROACH TO RAISING MICROBIOLOGICAL RESISTANCE OF MATERIALS AND ARTICLES. INVESTIGATION TECHNIQUES FOR STUDYING MICROBIOLOGICAL RESISTANCE OF *materials and efficiency of protection facilities*

Development of protection measures is desirable at all stages of articles lifetime (design, production and operation). Some investigations can be performed in the frames of the general system approach. Figure 7.1 shows components of this approach.

The first stage of proceedings is analysis of construction, process and operation features of the article affecting its microbiological resistance. Hence, the information about the microbiological resistance of the article, materials used for its production, active media, parts, aggregates, systems, protection measures and modes, as well as microorganisms-destructors is considered. Data on the effect of applied geometrical shapes, used assembling, stagnation zones, airproof volumes, as well as information about external factors of operation affecting microbiological damage are analyzed.

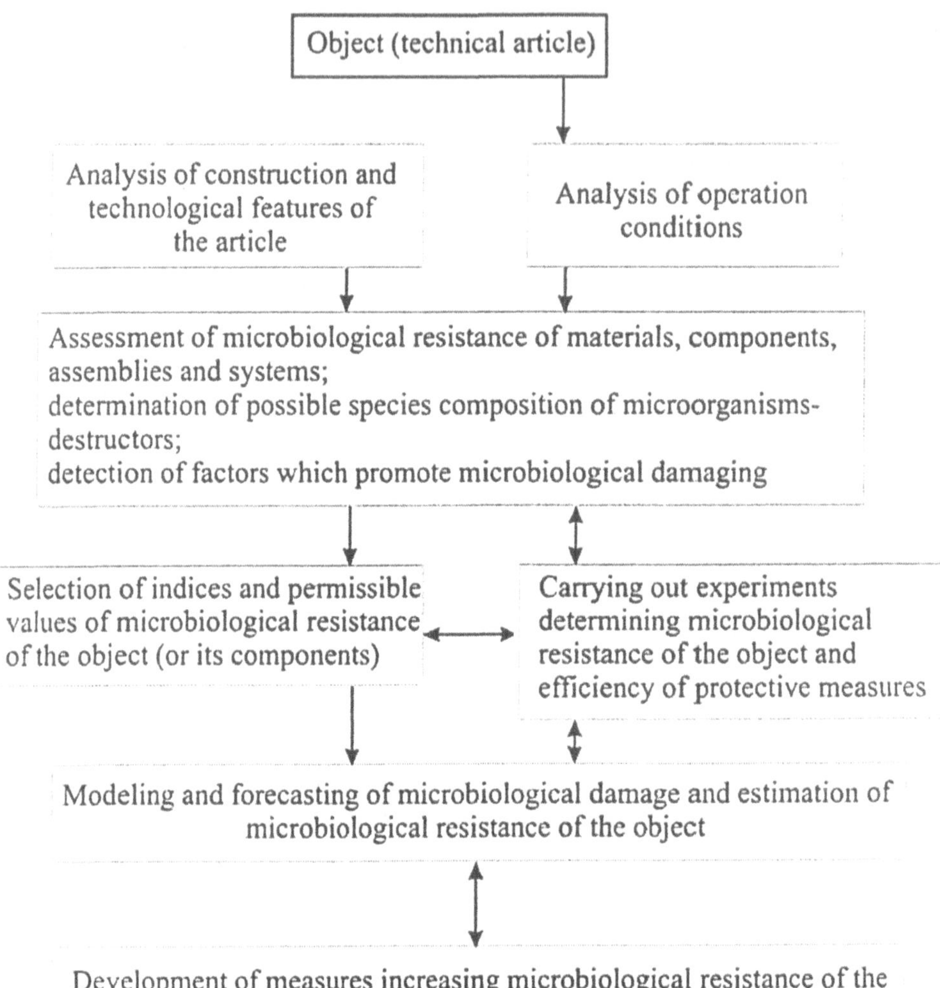

Figure 7.1. The scheme of system approach to the increase of microbiological resistance of technical facilities

As this stage is completed, objects of the microbiological damage (materials, parts, etc.), species composition of microorganisms-destructors, main factors of the external medium promoting interaction between microorganisms and materials are determined.

The next important stage of the work is selection of microbiological resistance indices and their assumable values based on the results of special investigations with regard to functional designation of the article (or its components) and some known (given) period of time, τ (operation period, inter-repair period, frequency of servicing, etc.).

Characteristics of microorganism growth on the surface (or in the volume) of a material (biomass, m) and/or changes in properties of materials (property index, A), as well as rates (intensities) of the process proceeding are usually used as indices of the microbiological resistance.

Assumable values of biomass (m_{ass}) and property variation (A_{ass}) are set on the basis of threshold values of these indices, exceeding of which causes unfavorable changes in the object conditions. For electric articles, the assumable index of resistance to fungi is growth of fungi estimated by degree 3. For optical parts, it is degree 2 [2, 4].

The stage of modeling and forecasting of microbiological damages of an article (or its components) is carried out regarding construction and process features and operation factors, which cause the determining effect on interaction between microorganisms and materials.

Basing on the data existing, one may suggest the use of various approaches for modeling and forecasting results of the mentioned interaction (expert estimation, physical and mathematical modeling, etc.).

Microbiological resistance of the articles is estimated by comparing indices obtained by modeling (forecasting) with their assumable values. Non-compliance of the condition $m(\tau) \leq m_{ass}$ at $A(\tau) \leq A_{ass}$ testifies about insufficient microbiological resistance of the object (its component) and gives the basis for a decision about necessary modification of protection measures and methods.

Measures increasing microbiological resistance are developed by analyzing information about the existing protection measures and methods and based on technical and economic expediency of their application. The information about existing protection measures and methods is generalized in refs. [2, 4, 180].

Generally, microbiological damaging may be inhibited with the help of four groups of effects as follows:

The effect on medium – changing the composition and parameters, including concentration of additives favorable for microorganisms and possible contamination, temperature and humidity;

The effect on microorganisms – changing their species composition and quantity in the medium, injection of substances inhibiting metabolism of microbial cells into the medium, removing or annihilation of microorganisms-destructors present on the surface (in the volume) of materials and components;

The effect on a technical facility – rational designing which prevents appearance and development of microbiological damages, object cleaning from contaminants promoting growth of microorganisms;

The complex effect representing various combinations of the above-mentioned groups.

The existing protection measures and methods may be classified with regard to their effect on the impact of microorganisms on the material inducing its damage. Protection can be achieved at every stage of the process discussed in Section 1.2.

At the stage of spreading and transfer of microorganisms, protection can be performed by selection of the operation regions, which minimize possibility of object contamination by microorganisms-destructors; sterilization of air flows; partial or full isolation of the object from contamination by microorganisms.

Adhesion may be regulated by changing hydrophilic-hydrophobic balance and water absorption, inhibiting cellular processes of metabolites-adhesives production, changing roughness of the surface and external conditions that minimizes adhesion forces.

Intensity of the microbiological growth is reduced by reducing concentration of the components nutritious for microorganisms in the material, introduction of substances inhibiting metabolism in cells (biocides), creation of external conditions unfavorable for microorganisms' growth.

Variations in material properties induced by microorganisms are regulated by changing the material structure (increase of crystallinity degree and orientation of polymers, macromolecule cross-linking with formation of structural networks), changing chemical structure (introduction of substituents into polymer macromolecules, which hinder approach of aggressive metabolites to chemically unstable bonds), changing the material composition (by introduction of mineral fillers capable of diffusion to the surface and formation of a protective layer at the interface), modifying the material surface by formation of an isolating layer (layers) on it possessing different physicochemical properties, resistant to metabolites, changing the type of surface layer stressed state of a material, a component, or an article, for example, by formation of residual compression stresses preventing material cracking in the presence of metabolites, changing external conditions in order to minimize rates of sorption and chemical (electrochemical) processes proceeding.

General directions of the protection mentioned above can be realized by creation and application of microbiologically resistant materials and structures, special protection facilities and methods, as well as adjustment of article operation and servicing conditions preventing microbiological damage.

The efficiency of developed measures is estimated with the help modeling and predicting their effect on microbiological resistance of the object. The sufficiency of any measure testifies fulfillment of the condition $m(\tau) \leq m_{ass}$ or $A(\tau) \leq A_{ass}$.

Note that performance of the above-mentioned stages of investigations is based on data on the microbiological resistance of materials, articles, protection facilities and methods, and forecasting of change in the resistance under conditions of operation of a technical facility. If the data are absent, they are obtained by performing a complex of laboratory or full-scale tests.

The technique for detection of biodamages is discussed in Chapter 3. It suggests application of a complex of various microbiological, biochemical and physicochemical methods of determining parameters and indices of the microbiological damage. The possibility and intensity of biodamage proceeding is concluded from the analysis of the totality of data obtained during all recommended investigations. The methodology can be used for carrying out target inspections, investigations of technical conditions of the facilities, including emergency and failed articles.

The determination techniques for quantitative indices of microbiological damaging of materials and efficiency of protection measures are discussed in detail in Chapters 4, 5, and 6 and in two State Standards (GOSTs), which describe determination of indices of separate stages of microorganism damaging effect, adhesion and growth of microbial cells on materials and changes in their properties. The techniques suggest experimental obtaining of dependencies of quantitative parameters of these stages on time of object contact (including biocide substances) with microorganisms. Using data from experiments and appropriate analytical models, indices of the mentioned processes (stages) are calculated, values of which characterize microbiological resistance of materials: their ability to be changed under the effect (during a given time) of biodestructor, to adhere microbial cells, and to provide their growth and change their properties in contact with microorganisms.

The laboratory investigation method of adhesion of microscopic fungi to polymeric materials provides obtaining values of the adhesion force, F_{∞}^{50}, and the rate constants of this force formation, K.

A generalized adhesion index (Δ) can also be calculated, which considers both kinetic component and the adhesion force value of spores to the surfaces. The index Δ equals the area beneath the kinetic curve $F_{\infty}^{50} = f(t)$. As for many microorganism-material systems the time of formation completion of adhesive interaction is shorter than 36 hours, the generalized index is preferably calculated for the time period of 0 – 36 hours, wherefrom:

$$\Delta = \int_{t_0}^{t_1} f\left(F_t^{50}\right)dt = \int_{t_0=0}^{t_1=36} F_t^{50}\left[1-\exp(-Kt)\right]dt = F_t^{50}\left[36+K^{-1}\exp(-36K)-K^{-1}\right].$$

The greater the index Δ is, the higher the material ability to adhesion is.

The determination method of quantitative indices of microorganism growth and change in properties of polymeric materials is described in GOST 9.049-91 and GOST 9.803-88. Experimental obtaining of the following values can be regulated: duration of the microorganism adaptation period to the material, L; maximal value of controlled growth parameter, A_{∞}; specific growth rate, b. Quantitative parameters controlled during the experiment are: dry specific biomass (gravimetric method) and intensity of β-radiation (isotope method).

The method determining biocide properties of substances, standardized in GOST 9.803-88 (Chapter 5), is analogous to the above-described tests on material resistance to microbiological growth. The growth is characterized by the dry specific biomass of microscopic fungi, formed during their cultivation on a hydrogel support contacting with a liquid nutritious medium in the presence of studied chemical compounds of various concentrations.

Experimentally, a series of kinetic dependencies of the specific biomass of fungi at various concentrations of the studied compound are deduced, and growth parameters (L and b) are calculated. Then using the equation of growth inhibition (Chapter 5) associating duration of the lag-phase, L, and the specific growth rate, b, with the fungicide concentration, constants of these equations, K_L and K_b, are calculated. Values of these constants characterize the fungicide efficiency. The lower is K_b and the higher is K_L, the more efficient the fungicide is.

7.2. MATHEMATICAL MODELING AND PROGNOSIS OF MICROBIOLOGICAL DAMAGING OF MATERIALS

The ideas developed in the monograph allowed development of mathematical models for prognosis of microbiological damage by extrapolation methods of the existing information about the process and mathematical modeling [357].

The extrapolation method is based on application of simple mathematical equations associating forecasted value with various indices of a process proceeding with time. Mathematical modeling is carried out using a totality of dependencies, which describe the structure of processes with regard to the effect of random factors on them.

These two groups of methods are characterized by criteria of material damage induced by microorganisms. They are: 1 – the quantity of biomass on the material surface (m), corresponded to its assumable value (m_{ass}), or the same value of the biomass, at which properties of the material begin changing; 2 – a change of a property index of the material (A), induced be a biofactor, to a critical (selected) level (A_{ass}).

Prediction of occurrence of a microbiological damage is reduced to determination of time (τ_{dam}) of biomass $m(\tau_{dam})$ the assumable value m_{ass}, or τ_{dam} necessary for changing a property index $A(\tau_{dam})$ to a critical level A_{ass}:

$$m(\tau_{dam}) \geq m_{ass} \qquad (7.1)$$
$$A(\tau_{dam}) \geq A_{ass}.$$

7.2.1. Extrapolation of the information about the process

Shown in Chapters 4 – 6 are kinetic equations of every discussed stage of microbiological damaging of the material and equations associating quantitative indices of these stages with temperature and humidity of the environment. In the aggregate, these equations represent the model of the process. They allow prediction of the intensity of its proceeding in any stage of microorganism interaction with materials. Principles of predictions are well-known and based on the rule of temperature-time or humidity-time analogy [4]. The presence of, at least, two experimental kinetic curves of variations in predicted parameter, obtained at various temperatures (T) and air humidity (), is suggested. From these experimental data and analytical equations describing stages of the process, its indices are calculated at T and φ used in the experiments. Then parameters of dependencies associating process indices with temperature and humidity are calculated from the obtained values. Finally, on the basis of the above-mentioned indices and parameters and using appropriate kinetic equations, a change of predicted values at any time and any combination of parameters T and φ is determined.

Note that prediction dependencies suggested in Chapters 4 – 6 are obtained first for various estimations and determination of the process origin of microorganism interaction with materials. Parameters (indices) of these dependencies possess a definite physical meaning and reflect the features of appropriate stages of microbiological damaging. At the same time, formal models and indices [4] may also be applied to solve tasks of forecasting by extrapolation. Composition of such models often reduces the volume of experimental studies and simplifies appropriate calculations.

Simple regression expressions were used by the authors in modeling and prediction of initial stages of microscopic fungi development on onboard electric lines' wire insulation from varnished fabric under real conditions of technical facility operations.

The analysis of kinetic data, carried out in Chapter 6, shows that microbiological damage may be characterized by the biomass value (m_{ass}) equal ~0.1 mg/cm^2. Above this value, electrical resistance of wires with insulation from varnished fabric is changed in contact with microscopic fungi independently of the temperature and humid conditions. In this case, prediction is reduced to determination of time, τ_{dam}, necessary for accumulation of the mentioned quantity of biomass, corresponded to the criterion selected.

Time, τ_{dam}, may be represented as a sum of time necessary for microorganism adaptation to the material, L, and time for biomass growing up to m_{ass} value, τ_m.

As shown in Chapter 5, the value of L in the humidity range of 75 – 100% (at T = +15 - +29°C) is generally determined by temperature of microorganism cultivation on the material. This allows application of the following equation to determination of the lag-phase (L_T) at constant temperature:

$$L_T = B_{1L} \cdot T + B_{1L} \cdot T^Z + A_L, \qquad (7.2)$$

where B_{1L} and B_{2L} are coefficients; A_L is a constant; Z is the exponent.

Values of A_L, B_{1L}, B_{2L} and Z were determined by the main component method (the PCRG method, PPSA package). The following values have been obtained: B_{1L} = -6.0491·10^2; B_{2L} = 2.1448·10^2; Z = 1.25; A_L = 3.2333·10^3. Coefficients of multiple correlation and determination equal 1.0, and the mean-square uncertainty is 3.3222·10^{-3} that testifies about acceptability of suggested regression equation to prediction of the L_T value.

To analyze dependencies of the lag-phase (Chapter 4), the authors have used a supposition that microbial cells accumulate some biologically active component during adaptation to the material. Let us assume that the biomass growth is initiated at a concentration of this substance in cells, equal M. In accordance with this hypothesis, in the first approximation it has been assumed that the time dependence of biologically active component concentration increase is linear and, similar to L_T, is determined first by temperature of the environment:

$$M = V_T^L \times L_T, \tag{7.3}$$

where V_T^L is the rate of biologically active component concentration increase at constant temperature, T.

Variation of the component M_{T_i} concentration during time τ_{T_i}, characterized by constant temperature T_i, may be expressed by the following dependence:

$$M_{T_i} = V_{T_i}^L \times \tau_{T_i}. \tag{7.4}$$

Then at the end of microorganism adaptation to the material, the quantity of the component M may be presented as the sum of M_{T_i} values for n time periods with constant temperatures:

$$M = \sum_{i=1}^{n} M_{T_i} = \sum_{i=1}^{n} V_{T_i}^L \cdot \tau_{T_i} = V_{T_i}^L \cdot L_{T_i}. \tag{7.5}$$

Using equation (7.5), one may write down the condition for the lag-phase end:

$$\sum_{i=1}^{n} \frac{V_{T_i}^L \cdot \tau_{T_i}}{V_{T_i}^L \cdot L_{T_i}} = 1. \tag{7.6}$$

The equations (7.2) and (7.6) allow calculation of the lag-phase duration under conditions of variable temperature and humidity of the environment.

In the areas of biomass growth kinetic curves after the lag-phase end and before reaching m equal 0.1 mg/cm^2, the dependence $m = f(\tau)$ was suggested to be linear and described by the following expression:

$$m_{T,\varphi} = V_{T,\varphi}^m \times \tau_{T,\varphi}, \tag{7.7}$$

where $V_{T,\varphi}^m$ is the biomass growth rate at constant temperature and air humidity; $\tau_{T,\varphi}$ is the time of biomass growth at constant temperature and air humidity.

For different combinations of temperature and air humidity, the value of $V_{T,\varphi}^m$ was determined by a regressive equation of the following type:

$$V^m_{T,\varphi} = B_{1m} \times \varphi + B_{2m} \times T + B_{3m} \times \varphi^2 + B_{4m} \times T^2 + A_m, \quad (7.8)$$

where B_{1m}, B_{2m}, B_{3m}, and B_{4m} are coefficients; A_m is a constant.

Values of coefficients for experimental data shown in Chapter 4 are the following: $B_{1m} = 5.0938\times10^{-4}$; $B_{2m} = 4.7408\times10^{-4}$; $B_{3m} = -2.8982\times10^{-6}$; $B_{4m} = 1.4291\times10^{-4}$; $A_m = -2.4798\times10^{-2}$. The regression and determination coefficients equaled 0.96 and 0.92, respectively, which indicates accuracy high enough for determination of the biomass growth rate by the regression equation suggested.

To take into account temperature and humidity conditions varying during the microorganism cultivation, a method has been used, analogous to determination of the lag-phase duration. A colony biomass (m) at a moment (τ) may be presented by an expression of the following type:

$$m = \sum_{i=1}^{n} m_{T_i\varphi_i} = \sum_{i=1}^{n} \tau_{T_i\varphi_i} \cdot V^m_{T_i\varphi_i}, \quad (7.9)$$

where $V^m_{T_i\varphi_i}$ is the biomass growth rate at constant T_i and φ_i; $\tau_{T_i\varphi_i}$ is the duration of the i-th period with constant T_i and φ_i values; n is the number of periods with constant temperature and humidity conditions.

The expression obtained allows prediction of time (τ_m) of biomass reaching the critical value (0.1 mg/cm^2) under any varying temperature and humidity conditions. The values of τ_{T_i} and $\tau_{T_i\varphi_i}$ may be selected on the basis of analysis of the data on temperature and humidity of particular climatic region of the technical facility operation.

The suggested predicting model was checked on the basis of study of the biomass growth kinetics on electric wire insulation from varnished fabric under arbitrary temperature and humidity mode varied during incubation of the wire samples, infected by microscopic fungi. The tests were terminated, when the colony biomass reached ~0.1 mg/cm^2 density.

Equations (7.2) and (7.8) were used in calculations of the lag-phases and biomass growth rates for every period under constant temperature and humidity conditions. After that, using conditions (7.6) and (7.9), L and τ_m values were determined in conditions of T and φ varied during cultivation. Thereafter, the time during which the system reaches the critical state (m_{ass}) was calculated: $\tau_{dam} = L + \tau_m$. Table 7.1 shows the calculation results.

Table 7.1

Experimental and calculated values of the lag-phase and microorganism growth rates

Test mode		$\tau_{T_i\varphi_i}$	$L_{T_i\varphi_i}$, hour		L, hour		$V^m_{T_i\varphi_i}$ ×10³, (mg/cm²)/h		τ_m, hour		τ_{dam}, hour	
T, °C	φ, %		Exp.	Calc.	Exp.	Calc.	Exp.	Calc.	Exp.	Calc.	Exp.	Calc.
20	100	35	168	192	117	135	No biomass growth is observed					
29	75	82	110	115								
29	75	19	—	—	—	—	20	22	76	59	193	194
15	75	57	—	—	—	—	12	14				

Clearly that every constant combination of φ and T gives equal experimental and calculated values of the lag-phase, and the biomass growth rates, as well as satisfactory convergence is observed for the appropriate values of L, τ_m and τ_{dam}, determined in an experiment or calculated by an appropriate method. These results prove authenticity of the prediction model developed. It gives a possibility to predict development of microbiological damage of materials at the early stage based on laboratory tests of material resistance to the effect of microorganisms and statistical data on temperature and humidity complex of the atmosphere.

Application of statistical extrapolation methods suggests the determinate basis of the microbiological damaging. Thus, according to the opinion of many scientists [2, 4, 25, 59], regularities of the effect of the whole variety of factors on microorganism interaction with abiotic objects (materials, articles) cannot be considered or cleared out at present. Extrapolation models used consider parameters of temperature and humidity complex of the atmosphere only. Expansion of the number of sufficient variable factors in these models is very difficult, because quantitative information about the environment influence on the considered process is short.

Moreover, the form of extrapolation equation in our investigations is most often determined by approximation of experimental data by a series of various functions. Hence, the function is used, which displays the lowest approximation uncertainty. Clearly such approach allows proof of suitability of the selected equation for description of the process proceeding in a limited prediction area.

The stochastic type of microorganism interaction with materials gives an opportunity to suggest that the mathematical modeling is the most perspective method of predicting microbiological degradation.

7.2.2. Mathematical modeling of the process

Quantitative descriptions of the microbiological degradation regularities, described in Chapters 4 – 6, are based on the phenomenological approach to the interaction between microorganisms-degraders and materials. Many analytical models suggested are of empirical or semi-empirical type and are unable to discover the essence of physical and chemical phenomena, which stipulate these interactions. Based on such phenomenological information, it seems better to develop a forecasting mathematical model with the help of differential equations [357]. The stage-by-stage structure of the microbiological damaging, discussed in the previous sections, may be described by a system of differential equations with equitable dynamic variables, which are: concentration of microbial cells or their biomass (for the stages of cell transport from the environment, their adhesion and growth on the material) and material properties (for the stage of property degradation).

As a rule, under real conditions of technical object operation, microorganisms are transported to materials and articles by air from the surrounding atmosphere. That is why realization of the microorganism-material interaction needs the presence of viable cells of microorganisms-degraders in the surface layer of the atmosphere. It is known (Chapter 1) that the concentration of such cells in the air, $C_a(\tau)$, is not constant. It depends on the season, daytime, type of soils, flora composition and dislocation, locality relief, meteorological conditions and other factors.

In the place of technical object location at the given moment of time, τ, $C_a(\tau)$ may be calculated from data on the average concentration of microbial cells in the air during the day round, $C_k(\tau)$, where k is the index of the month since the beginning of the article operation. Hence, daily oscillations of microbial cell concentrations are taken into account in the average time of the cell occurrence, τ_{co}, and presence, τ_{cp}, in the atmosphere. Thus, $C_a(\tau)$ will be calculated from the following expression:

$$C_a(\tau) = C_k(\tau)\cdot\delta_{kc}, \tag{7.10}$$

$$\delta_{kc} = \begin{cases} 1, \text{if } 8{,}760\cdot i + \tau_{co} + 24\cdot l + \tau_{k-1} \le \tau \le 8{,}760\cdot i + \tau_{co} + \tau_{cp} + 24\cdot l + \tau_{k-1} \\ 0, \text{for other values of } \tau \end{cases};$$

$$l = \frac{\tau - \tau_{k-1} - 8{,}760\cdot i}{24},\ i = \frac{\tau}{8{,}760},$$

where $C_k(\tau)$ is the average volumetric concentration of viable cells of microorganisms-degraders in the air during one day in the k-th month; i is the index related to the year of the article (material) operation; l is the number of the day; τ_k is the time corresponded to the end of the k-th month ($k \leq 1, 2, \ldots, 12$); τ_{co} and τ_{cp} are the average times of the microorganism-degrader cell occurrence and presence in the atmosphere during a day of the k-th month, respectively; 8,760 is the year duration in hours.

$C_a(\tau)$ is calculated based on the results of direct experimental determination of the microorganism-degrader cell concentration under particular conditions, in which the technical object is present. For this purpose, the known analytical methods determining microbiological contamination of air are applied (Chapter 1).

In accordance with the stage-by-stage approach used, the first stage of microbiological damaging is transport (supply) of microbial cells from the environment to the material surface and their adhesion on it. The time dependence of the cell concentration in the zone of adhesive force action at the material surface, $C_{ms}(\tau)$, may be presented by a differential equation as follows:

$$\frac{dC_{ms}(\tau)}{d\tau} = C_a(\tau) \cdot k_1(\tau) \cdot q_1(\tau). \tag{7.11}$$

This equation is solved as follows:

$$C_{ms}(\tau) = C_{ms}(\tau - \Delta\tau) + C_a(\tau) \cdot k_1(\tau) \cdot q_1(\tau) \cdot \Delta\tau,$$

where $k_1(\tau)$ is the coefficient describing the cell transport rate to the zone of active adhesive forces at the material surface; $q_1(\tau)$ is the transport coefficient describing an "equilibrium" state of the microorganism-material system under current conditions and the moment of time and reflecting the relation between possible quantity of cells in the zone of active adhesive forces, $C_{PM}^{\max}$, and their quantity in the environment, $C_a(\tau)$.

Values of $k_1(\tau)$ and $q_1(\tau)$ are determined experimentally using a special chamber, the main parameters of which are shown in Chapter 2, or a wind tunnel providing constant test conditions with time (Chapter 1).

Microbial cells present in the zone of active adhesive forces are fixed (adhered) to the material surface. Concentration of the cells adhered by the material, $C_z(\tau)$, can be expressed by the equation:

$$\frac{dC_z(\tau)}{d\tau} = C_{PM}(\tau)\cdot k_2(\tau)\cdot q_2(\tau), \qquad (7.12)$$

which is solved as follows:

$$C_z(\tau) = C_z(\tau - \Delta\tau) + C_{PM}(\tau)\cdot k_2(\tau)\cdot q_2(\tau)\cdot\Delta\tau,$$

where $k_2(\tau)$ is the coefficient characterizing the rate of the adhesive force formation; $q_2(\tau)$ is the adhesion coefficient.

Values of $k_2(\tau)$ and $q_2(\tau)$ are determined experimentally by the method of centrifugal detachment (Chapter 4).

The development of microbial cells adhered to the material (article) and the biomass growth (the second stage of the microbiological damage) are initiated under a complex of environmental conditions, Q_{raz} (temperature, humidity, etc.), favorable for the microorganisms.

Environmental parameters stipulating possibility of the microorganism development are determined experimentally. For the factors such as temperature, air humidity and concentration of microbial cells adhered to the material, condition of the microorganism-degrader development may be presented as follows:

$$Q_{raz} = \begin{cases} 1, \text{if } \varphi(\tau) \geq \varphi_{raz}^{\min}(\tau), T_{raz}^{\max}(\tau) \geq T(\tau) \geq T_{raz}^{\min}(\tau), C_z(\tau) \geq C_{z\,raz}(\tau) \\ 0, \text{at other values} \end{cases}, \qquad (7.13)$$

where $\varphi(\tau)$ and $T(\tau)$ are air humidity and temperature, respectively; $\varphi_{raz}^{\min}(\tau)$ is the minimal air humidity providing development of the microbial cells at temperature $T(\tau)$; $T_{raz}^{\max}(\tau)$ and $T_{raz}^{\min}(\tau)$ are the maximal and minimal temperatures, within which range microorganisms may develop (at air humidity φ); $C_{z\,raz}(\tau)$ in the concentration of cells adhered to the material providing the beginning of colony development under the given temperature and humidity conditions.

When the conditions $Q_{raz}(\tau) = 1$ is fulfilled, the microorganism begins developing on the material. Biomass at any given moment of time, $m(\tau)$, may be deduced from the following differential equation:

at $Q_{raz}(\tau) = 1$,

$$\frac{dm(\tau)}{d\tau} = C_z(\tau)\cdot m_{cell}\cdot k_z(\tau)\cdot q_z(\tau), \qquad (7.14)$$

solution of which is $m(\tau) = m(\tau - \Delta\tau) + C_z(\tau)\cdot m_{cell}\cdot k_z(\tau)\cdot q_z(\tau)\cdot\Delta\tau$, and

at $Q_{raz}(\tau) = 0$,

$$m(\tau) = 0, \tag{7.15}$$

where m_{cell} is the mass of the microbial cell; $k_z(\tau)$ is the coefficient characterizing the rate of the biomass growth; $q_z(\tau)$ is the coefficient of the biomass growth considering its possible increase in relation to the initial value ($C_z m_{cell}$) under current environmental conditions and at $C_z(\tau)$ value.

Values of coefficients $k_z(\tau)$ and $q_z(\tau)$ are obtained in accordance with the technique discussed in Chapter 5.

The contact with the biomass is accompanied by variations of material properties. Dependence of any property index $A(\tau)$ on duration of the microorganism impact on the material may be deduced from the following equation:

at $Q_{raz}(\tau) = 1$,

$$\frac{dA(\tau)}{d\tau} = A_0 \cdot k_4(\tau) \cdot q_4(\tau), \tag{7.16}$$

solution of which is $A(\tau) = A(\tau - \Delta\tau) \pm A_0\cdot m_{cell}\cdot k_4(\tau)\cdot q_4(\tau)\cdot\Delta\tau$,

at $Q_{raz}(\tau) = 0$,

$$A(\tau) = A_0, \tag{7.17}$$

where A_0 is the initial value of the material property; $k_4(\tau)$ is the coefficient characterizing the rate of microbiological degradation of the property; $q_4(\tau)$ is the coefficient of microbiological degradation of the property, which characterizes possible effect of variation in the property for the given microorganism-material couple under current test conditions.

Values of coefficients $k_z(\tau)$ and $q_z(\tau)$ are obtained in laboratory tests described in Chapter 6.

Both suggested differential equations and analytical models and the appropriate indices of microbiological degradation stages (adhesion, growth, or property variation) may be used in the mathematic model.

During solving differential equations, real conditions are simulated by modeling random impacts of environmental factors at any given moment of time in particular areas of technical object location. In accordance with refs. [357 - 360], environmental factors are presented in the form of stationary random processes, statistic parameters of which are calculated from multi-year meteorological data. The equations modeling random effects of temperature, humidity, air flow speed and atmospheric fallout in the form of rain or fog are discussed in refs. [358 – 360].

Using biomass values, m_{ass}, as the damage criteria, time of its occurrence on the object surface (time of damage occurrence, τ_{surf}) is determined by linear interpolation of modeling results by the argument, τ.

$$\tau_{surf} = \tau + \frac{m_{ass} + m(\tau)}{m(\tau + \Delta\tau) - m(\tau)} \cdot [\tau - (\tau + \Delta\tau)]. \tag{7.18}$$

In analog with (7.18), the period duration during which impact of microorganisms induces a variation of the property index to a level A_{ass} is calculated by the relation:

$$\tau_{surf} = \tau + \frac{A_{ass} + A(\tau)}{A(\tau + \Delta\tau) - A(\tau)} \cdot [\tau - (\tau + \Delta\tau)]. \tag{7.19}$$

Hence, time of occurrence of the microbiological damage of the material is forecasted by integrating differential equations with regard to an accidental impact of environmental factors until conditions (7.1) are fulfilled:

$$m_{ass} - m(\tau) > 0,\; m_{ass} - m(\tau + \Delta\tau) < 0$$

or

$$A_{ass} - A(\tau) > 0,\; A_{ass} - A(\tau + \Delta\tau) < 0.$$

The initial values of climatic and biological factors as well as coefficients of microorganism interaction with materials are given in tables with the number of entries corresponded to the number of arguments. Intermediate values from the tables are chosen with the help of linear interpolation by every argument.

Reliability of forecasting, obtained with the help of the developed model, was checked on the basis of comparative estimation of the calculations and laboratory test data on electric wires with insulation from varnished fabric. The wires were exposed in a special chamber under temperature (25 - 29°C), humidity (75 - 100%) and the absence of air flows, which is favorable for development of *Aspergillus niger*. The microscopic fungus spore concentration in the test chamber varied during the day from 0 to 20 – 30 spore/m^3. The criterion of biodamages was biomass (m_{ass}) equal 0.1 mg/cm^2, at which dielectric properties of the insulation are changed (Chapter 6). It was been found experimentally that the biomass, m_{ass}, is formed on the wires after 85 – 117 days of tests. Calculations carried out by equations (7.10) – (7.15) using the above-shown initial data have indicated that the damage may be expected 98 days after. The calculation uncertainty is below 20%.

Hence, a mathematical model of microbiological damage of materials is developed, which represents the system of differential equations with dynamic variables. It considers the process structure, environmental impacts on it, and forecasts time of occurrence and intensity of the process development under particular conditions of technical object operation stage by stage.

Microbiological damage models obtained are used by research institutions during the investigations aimed at protection of technical facilities against biodamages.

7.3. METHODS AND FACILITIES FOR PROTECTION OF ARTICLES AGAINST MICROBIOLOGICAL DAMAGES

The investigations of microbiological damages of technical facilities under operation conditions, species composition and the features of microorganisms-destructors of the materials allow formulation of suggestions to maintenance of the required biological resistance of the articles. They are realized in a series of scientific and technical achievements, which regulate technical requirements to articles and protection facilities.

OTT and GOST 24627 system documents contain requirements to construction, selection of protection materials and facilities, article maintenance volume and regularity, aimed at prevention and elimination of microbiological damages with regard to the effect of their consequences on the efficiency of components, aggregates and systems of articles. These documents are applied by various institutions to stating requirements to particular types of machinery units, as well as by the industrial developers of these units.

Requirements to protection measures regulate the use of microorganism strains, separated from materials damaged under operation conditions, as the test-cultures for assessment of these measures. Such strains, stored alive in a specialized collection of microorganism strains, most fully reflect the features of microorganism impacts on particular materials under real operation conditions.

Table 7.2 lists protection facilities created in cooperation with various industrial institutions and their brief characteristics. The compositions of the protection measures were developed with regard to rational combination of protective abilities to the impact of microorganisms and climatic factors, manufacturability of production and application, ecological and toxicological loading, etc. Assessment of the protection efficiency against the impact of microorganisms and appropriate selection of required compositions were carried out using quantitative determination methods for the microbiological resistance of materials and biocide properties of chemical compositions (refer to Section 7.1).

At present, biocide substances (B-1779 bactericide and F-2223 fungicide) are belonged to the group of the most perspective, the so-called ecological-biochemical protectors of technical facilities. B-1779 bactericide represents lyophilized inactivated biomass of *Bacillus sp.* strain, and F-2223 fungicide is an individual chemical compound extracted from mycelium of *Tolypocladium sp.* microscopic fungus, strain 2223. Laboratory samples of some compositions of these substances possess rather high biocide activity.

Other substances listed in Table 7.2 are industrially produced. Eight separate issues of methodical guides on their application have been compiled. Application of MIKON and IVVS-1-94 substances is standardized in GOST 9.014-78 and GOST 9.014-80.

Table 7.2

The list and characterization of substances protecting technical facilities and machinery from microbiological damages

Title, documents regulating application, and institution-manufacturer	Composition, biocide component, concentration (wt.%)	Designation	Application method
Film-forming protective substances			
1. **IVVS-1-94** TU 0255-234-00151526-94 GOST VD 9.014-80 (changes No. 5 dated July 01, 1998) VNIKTI oil and chemical equipment	Dispersion of ceresin solution in white spirit, modifying additives and biocide in aqueous ammonium salts of synthetic fatty acids. **Biocide**: potassium bichromate – 0.3 – 0.4.	Protection from microbiological damages and aging of components from polymers, fabrics, leather, rubber-technical articles, varnishes and paints, package paper, cardboard, and wood. Protection of metals and alloys from atmospheric and microbiological corrosion. Suggested to	Film application by sprayers, paint-brushes, or dipping. Drying at temperature above +5°C during 0.5 – 2 hours. Composite consumption – 0.05 – 0.1 kg/m^2. Protective film thickness – 30 – 70 μm.

		substitute aqueous-wax ZVVD and IVVS-F compositions.	
2. Substance for chemical modification of optical parts surfaces. RF Patent 2,111,182 [361] S.I. Vavilov VNC GOI	SKTNFG-10 liquid rubber solution (polymethyl-trifluoropropyl(vinylmethyl)-α,ω-dihydroxyloxane) and film-forming biocide in petroleum ether or toluene. **Biocide**: β-triethylstannylthio/ethyltrietoxysilane (TETS, stanilox) – 0.5 – 1.	Protection of optical parts (glasses, lenses, etc.) from adverse impact of microbiological and climatic factors of the environment.	Substance application by a cotton wool tampon or dipping. Drying at 15 - 70°C during 3 - 48 hours (with regard to temperature conditions). Substance consumption – 0.002 – 0.005 kg/m^2.
3. **Biokor-L** (Izol-1 composition) TU 0000-010-00209906-93 RF Patent No. 2,111,996 [362] VNIIKP, KB OR	Solution of vinyl chloride and vinyl acetate copolymer (VA-15), plasticizer (dioctylphthalate), antioxidant (2,2-bis/4-hydroxyphenyl/propane) and biocide in methyl ethyl ketone. **Biocide**: dichloromaleic acid N-cyclohexylimide (cymide) – 0.1 – 0.3.	Protection against microbiological damages and aging, atmosphere-induced corrosion of electric and radio articles: electric networks and internal assembly of components – electric wire and cable insulation from polyvinylchloride, polyethylene, and rubber; cambric, plug connector cases. Suggested to substitute KhV-114 (modified) and BIOKOR composites.	Coating application by pneumatic spraying, paint-brush or dipping. Drying at temperature above +5°C during 1 – 3 hours. Substance consumption – 0.008 – 0.015 kg/m of wire (13-mm section).
4. **VBO-91** TU 2290-015-00302287-94 RF Patent No. 2,091,528 [363]. CNIILKA	Water dispersion of 50% latex from butadiene-styrene rubber, a dye – biocide and alkaline agent. **Biocide**: 4,5,6-trichlorobenzoxazolon-2 (trilan) – 1 – 5.	Restoration of resistance to microbiological damages and aging of cellulose-containing fabrics (linen, semi-linen,	Fabric impregnation by paint sprayer or dipping at 0 - 35°C. Drying at 0 -

		cotton), lost during operation. Suggested to substitute KI-1 and VBO-86 composites.	120°C during 8 – 0.25 hours. Substance consumption – 0.09 – 0.25 kg/m^2 (due to fabric weight).
5. **Chemical modifier for material surface.** RF Patent No. 1,816,773 [364]. VAKhZ	Fluorine gas mixed with an inert gas.	Protection of polymeric materials from microbiological damages and negative impact of other environmental factors.	2 – 5 stages of treatment by the gas mixture. Treatment modes depend on the material type. Treatment time from 1 to 60 min, applied film thickness is 3 – 7 µm.
6. **Plasmachemical modifier for material surface.** RF Patent No. 1,684,084 [365].	Fluorokerosene vapor in high-frequency discharge plasma.	Protection of insulating polyvinylchloride plasticate, insulation varnished fabrics and rubbers based on butadiene rubber from microbiological damages and aging.	Exposure in bulky high-frequency discharge plasma in fluorokerosene vapor. Film thickness – several microns.
	Biocide agents		
7. **Volatile biocide – corrosion inhibitor VNK-LF-408** TU 301-03-135-94 RF Patent No. 2,083,719 [366] NPO Lenneftekhim	Product of morpholine and benzotriazoline condensation with benzaldehyde and biocide. **Biocide**: 1:1 mixture of citral (geranial) and eugenol ether oils – 0.05.	Protection of complicated articles from ferrous and non-ferrous metals, and various non-metal materials from microbiological damages and atmosphere-induced corrosion (at conservation of internal surfaces of hermetically closed volumes). Suggested to substitute VNK-L-20 and VNK-L-49 composites.	Applied in the form of powder filled up to a sublimate-former for obtaining inhibited air (consumption 150 g/m^3); inhibited paper (15 g/m^3); linapon (400 g/m^3); linasile and lingan (1,000 g/m^3); tabulin (100 g/m^3).

8. **Bactericide B-1777** RF Patent No. 2,141,523 [356] RF Patent No. 2,143,200 [367] NIIVS	Inactivated bacterial cell biomass of *Bacillus sp.* strain VKMV-1797 – 0.5 – 1.5 (in CIATIM-201 lubricant).	Protection of various materials, including technical oils and lubricants, against microbiological damages.	Injection to the composition of materials and protective substances, treatment of material surfaces.
9. **Fungicide F-2223.** GNCA	Extract from mycelium of *Tolypocladium sp.* microscopic fungus, strain No. 2223 – 0.01 – 0.03 (in AMS3)	Protection of various materials, including technical oils and lubricants, against microbiological damages.	Injection to the composition of materials and protective substances, treatment of material surfaces.
10. **Flamal 315A.** RF Patent No. 1,367,181 [368]	Individual substance – the liquid known as antipyrene (bis-0.01-isopropyl chlorobrimide-3-chlor-2-bromopropyl phosphanate) – 5 – 10.	Increase of efficiency of VBO-86 protective substance to restore resistance of articles from cellulose-containing fabrics to aging and microbiological damages.	Introduction to VBO-86 composition before treatment of fabrics (refer to par. 4 of the current Table).
11. **Fungistatic PTMI** RF Patent No. 1,350,855 [369]	Individual substance: N-n-tolyl maleimide – 0.05 (in ethanol solution).	Protection of metals and coatings.	Application of PTMI dissolved in ethanol to the surface by paint-brush or sprayer. Drying at room temperature during 20 – 30 min.
Detersives and disinfectants			
12. **Mikon** TU 24-041-0151489-93. RF Patent No. 2,084,498 [370] GOST 9.014-78. (Changes No. 5 dated July 01, 1998). NPO SintezPAV	Aqueous solution of triethanolamine, oleic acid, oxyphos, ethyl cellosolve; flotation reagent – oxale, carbamide and biocide. **Biocide**: poly(hexamethylene guanidine hydrochloride) (metacide) – 0.5 – 1.5.	Elimination of microbiological objects, process, operational, and oil and fatty contaminants from the material surface; surface disinfection. Suggested to substitute technical detersives: Vertolin-74, TMS-	Surface treatment by rags, brush, or special equipment for machinery treatment and cleaning. Active solution temperature is 20 - 80°C. Substance

		31, Impulse.	consumption: 0.15 – 0.35 kg/m^2.
13. **Detersive and disinfectant.** RF Patent No. 1,807,077 [371].	Technical detersive Impulse mixed with a biocide. Biocide: poly(hexamethylene guanidine hydrochloride) (metacide) – 0.4 – 0.8.	Elimination of microbiological objects, process, operational, and oil and fatty contaminants from the material surface; surface disinfection.	Surface treatment by rags, brush, or special equipment for machinery treatment and cleaning. Active solution temperature is 20 - 80°C. Substance consumption: 0.15 – 0.35 kg/m^2.

Compared with the known protective substances, the advantages of newly developed ones are as follows: they are designed to protect from a wide list of microorganisms-destructors (various species of microscopic fungi and bacteria); rather long periods of protective action (at least 5 years); high manufacturability of application under operation conditions; practically inexhaustible raw material and production base in the Russian Federation.

CONCLUSION

1. Ideas about damaging of materials by microorganisms as the process consisting of three main interconnected stages are developed. These processes are: adhesion of microorganisms, their growth on the material, and variation of material properties. Parameters and methods of their determination, which may indicate the presence and intensity of proceeding of each stage of the microbiological damage are proved experimentally. Analytical kinetic dependencies of these parameters reflecting the origin and interconnection of biodamaging stages, the effect of interacting objects properties and environmental conditions on it are determined. Parameters of the kinetic dependencies may be used as the indices of microbiological resistance of materials and efficiency of protective facilities.
2. A scientific methodological approach to studies of biodamages is suggested and experimentally stipulated. It suggests that the process consists of three basic stages of material interaction with a biodestructor and considering these interactions from positions of formal kinetics of the known physicochemical and biological processes: adhesion of finely dispersed particles, biological object development, and impact of liquid aggressive media on materials.
3. It is shown that at microscopic fungi-destructors precipitation in calm air, kinetics of changes in the quantity of adhered microorganisms subject to the precipitation equation of finely dispersed particles (the Stokes law) and is determined by radii of spores and the tilt of the material surface to the air flux. It is found that the adhesion force is increased with time of the spore contact with studied polymeric materials, varnish coatings or metals. This change follows the kinetic formula (the adhesion equation), which is general for various microorganism-material couples and environmental temperature and humidity conditions.
4. It is found that the increase of microscopic fungi and bacteria biomass on different polymeric materials, fuels and lubricants, as well as nutritious media containing biocidal substances under different temperature and humidity conditions follow the general kinetic equation (the growth equation). This equation describes kinetics of all phases of the microorganism growth on materials: adaptation and accelerated irregular, exponential and stationary phases.
5. Analytical dependencies connecting parameters of kinetic equations of adhesion and microorganism growth on materials with temperature, air

humidity, and biocide concentration (for the microorganism growth in a nutritious medium) have been obtained. Constants of these dependencies provide for quantitative assessment of the impact from these factors on biodamaging.

6. It is found that variations in operation properties induced by the impact of microscopic fungi and bacteria are associated with the effect of metabolites excreted by microorganisms and stipulated by both physical and chemical processes. Physical processes may induce both reversible and irreversible changes in properties. Chemical processes induce irreversible changes only.
7. It is shown that the reversible change in poly(methyl methacrylate) strength with regard to the growth stage of microorganisms contacting with it and stresses applied to the material is the result of metabolite absorption, because the latter reduces energy of new surface formation during deformation and (or) metabolite diffusion activated by the stress in the polymer volume. This diffusion induces local plasticization at the propagating crack tip. In the first case, the properties are changed with regard to regularities of the monomolecular adsorption of dissolved substances from solid surfaces; in the second case, they follow regularities typical of the hydrogen cracking of steels under mechanical stresses.

   The reversible change of PVC dielectric properties is induced by an increase of surface electric conductivity, caused by adsorption of metabolites and following regularities of monomolecular adsorption.
8. Reversible changes in varnished fabric electrical resistance induced by biomass are stipulated by sorption of metabolites in the material volume. Sorbed metabolites plasticize the varnished fabric making it more electrically resistant. Changes in electrical insulation resistance follow the regularities of liquid medium transfer in the material.
9. Irreversible increase of PVC-plasticate electrical resistance proceeding during a long-term impact of microorganisms is determined by the diffusive desorption of the plasticizer from the material. This process follows general sorption-diffusion regularities of low-molecular substance transfer from polymeric materials. The rate of it is limited by the plasticizer diffusion in the volume of PVC-plasticate.
10. Accumulation of the biomass in fuels and lubricants induces a significant increase of contents of mechanical admixtures (fuels and lubricants quality index). Changes in this index follow the regularities of the biomass increase.
11. Irreversible reduction of cotton thread and polysulfide U-30MES-5 sealer strength, and electrical resistance of varnished fabric is the result of chemical degradation, induced by the impact of metabolites, which changes chemical structure of the polymers. Properties of materials vary in accordance with the regularities typical of the polymer degradation in

chemically aggressive liquid media. The degradation regularities are inherent to the change in properties proceeding in the following kinetic areas of the process: the diffusion-kinetic area for cotton threads; the internal kinetic area for electrical resistance of varnished fabrics; the internal kinetic area for polysulfide sealer strength in the initial stage; the external diffusion-kinetic area for the final stage of the process.

12. Microbiological damage of D-16 aluminum alloy and St. 3 steel is induced by corrosion proceeding in metabolites according to the regularities of electrochemical corrosion of metals.
13. The investigation results allow development of techniques for detecting microbiological damages of operated materials and articles, determining quantitative indices of microbiological resistance of the materials, efficiency of protection facilities and methods, including estimation of biocidal properties of the substances, and mathematical modeling giving a possibility to predict biological resistance of the materials, efficiency of protection facilities and methods under particular conditions of the technical object operation.

Appendix 1.

The effect of various factors on damaging of polymeric materials by microorganisms

Analysis of data shown in the literature indicates the following factors of the environment seriously impacting the interaction between technophilic microorganisms and materials:

- Temperature and humidity of the atmospheric air and their oscillations;
- Air flow velocity;
- Solar radiation intensity;
- Type of vegetation cover and soils;
- Presence and composition of pollutions on the material surface;
- Geomagnetic field intensity.

Moreover, a possibility of occurrence of the microorganism-material interaction and its development intensity will mostly depend on activity and some other characteristics of the microorganisms.

It should also be taken into account that resistance to the microbiological damage is defined by the material origin. Structure, composition and properties are changed by the effect of the environment. Obviously, such aging of the material is also one of the factors defining its resistance to microbiological damage.

The factors seriously affecting occurrence and development of microbiological damage of onboard electric circuit insulation (OEC) of technical facilities have been determined using the method of expert estimations.

This method provides for polling specialists about the level of influence (significance) of some factors on the studied process. Based on the poll results and using test criteria of statistic hypotheses, coordination of opinions and reliable level of the factor significance have been determined.

Ten experts, specialists in the field of biodamages of materials and articles, have participated in the poll. The poll has been performed independently. In the initial stage, the specialists have chosen factors capable of affecting damaging of OEC wire insulation by microorganisms (Table A-1). Thereupon every specialist from the group has determined the significance level (rank) of factors conferring the rank 1 to the most significant and the rank 10 to the less significant ones. The rest of the factors have been conferred ranks from 2 to 9. This is the way of ranking the factors using the known algorithm and equations.

Table A1-1

Factors affecting occurrence and development of microbiological damages of OEC wire insulation

Factor	Factor number in the sequence	Conclusion about the factor significance
1. Temperature	1	Significant
2. Air humidity	2	Significant
3. Technogenic magnetic field	6	Significant
4. Humidity oscillations	7	Insignificant
5. Atmospheric fallout quantity	8	Insignificant
6. Air flows (wind)	9	Insignificant
7. Species composition and activity of microorganisms	3	Significant
8. Solar radiation	10	Insignificant
9. Wire insulation pollution	5	Significant
10. Insulation aging degree	4	Significant

Opinions of the specialists were checked by calculated (χ_c^2) and table (χ_t^2) values of the Pierson criterion and the concordance coefficient (W).

Calculated values (χ_c^2) were compared with the table (χ_t^2) ones (for the number of degrees of freedom, $\nu = K = 1$, and the significance, α). If $(\chi_c^2) > (\chi_t^2)$ and $0 < W < 1$, the hypothesis about non-random concordance of the specialist opinions is accepted.

The obtained value of W equaled 0.82, $(\chi_c^2) = 100.3$, and $(\chi_t^2) = 16.9$ at $\nu = 9$ and $\alpha = 0.05$, i.e. $(\chi_c^2) > (\chi_t^2)$. Hence, calculation performed has indicated that coordination in opinions of the experts is non-random.

Based on obtained average ranks, a regulated (in the order of significance) sequence of factors has been formulated (Table A1-1). Significant factors causing the highest effect on microbiological damaging of the insulation were determined with the help of the Student criterion (t_i, $i + 1$). Average ranks of the factors were compared in couples. In case if the obtained value of (t_i, $i + 1$) is smaller or equal to the table one, t_t ($\nu = 9$ and $\alpha = 0.05$), the difference of compared average ranks was considered insignificant and corresponded factors were assumed significant. The results obtained (Table A1-1) have allowed determination of the factors causing the most significant effect on occurrence and development of microbiological damaging of the electric wire insulation. In descending order of significance, such factors are temperature and air humidity, characteristics of microorganisms-biodestructors, the aging degree of the insulation, wire surface pollution, and intensity of technogenic magnetic field.

Obviously, the great importance for practical purposes is displayed by data on the type of influence of each above-mentioned factor on the insulation resistance to damage by microorganisms. The effect of various factors on the material degradability by microorganisms under real operation conditions was detected by comparing results of microbiological damages, detected in surveys of the technical conditions of articles and characteristics of considered factors. Microbiological damages were detected in accordance with the methodological scheme discussed in Chapter 3 of the present monograph. The technique for laboratory tests is shown in Chapter 2.

TEMPERATURE AND AIR HUMIDITY

Analysis of the data on microbiological damages of electric wire insulation of 30 articles indicates that the number of cases of damages is increased with the average monthly temperature, humidity and duration of periods combining temperature and humidity conditions ($T > 20°C$, $\varphi > 60\%$) favorable for growth of microorganisms.

The appropriate type of effect of the temperature-humidity complex on the growth of microorganisms on materials is doubt-free and taken into account by many authors. Quantitative aspects of these regularities are discussed in Chapters 4 and 5 of the present monograph on the examples of OEC electric wire insulation materials (varnished fabrics and PVC-plasticate).

THE SPECIES COMPOSITION AND ACTIVITY OF MICROORGANISMS-DESTRUCTORS

The species composition of microorganisms-destructors separated from electric wire insulation, operated in various climatic zones, is shown in Table A1-2.

Table A1-2

Species of microscopic fungi-destructors of OEC electric wire insulation in technical articles

Climatic region of operation (storage) of technical articles	Species of microorganisms-bioagents	Optimal temperatures of bioagent cultivation
Moderate	*Penicillium cyaneo-fulvum* Biourge	18 – 20
	Aspergillus niger	28 – 30
	Penicillium italicum Wehmer	20 – 23
Moderate-warm	*Penicillium velutinum* V. Beyma	25 – 28
	Penicillium raciborski Zaieski	25 – 28
	Aspergillus niger	28 – 30
Moderate-warm with non-frost winter	*Aspergillus niger*	28 – 30
	Aspergillus amstelodami Thom and Curch	25 – 32
Warm humid	*Penicillium capsulatum* Raper	27 – 29
	Aspergillus niger	28 – 30
	Aspergillus terreus var. aureus Thom and Raper	28 – 30
Hot dry	*Aspergillus niger*	28 – 30
	Aspergillus amstelodami Thom and Curch	25 – 32

Clearly microscopic fungi (micromycetes) are destructors of electric insulation, the most wide-spread in all surveyed climatic regions. Hence, every region is characterized by the self selection of micromycete species. However, *Aspergillus niger* microscopic fungus occupies electric wire insulation in all the climatic zones.

Intensity of development on the nutritious substrate and the range of temperatures optimal for vital activities are the important characteristics of microorganisms-destructors.

For microscopic fungi shown in Table A1-2, these characteristics have been determined from the studies of their growth kinetics in the Chapek-Dox medium at different temperatures. The colony development degree has been estimated by the increase of its radius (R). Strains separated by the authors from the wire insulation and analogous collection cultures from the Institute of Microbiology, RAS have been studied. It has been found that studied strains of fungi form colonies at temperatures above 15°C only. Typical dependencies of the micromycete colony radius on cultivation time (t) are shown in Figure A1-1. For all studied species of fungi and temperatures of cultivation, kinetic curves possess almost linear areas. Mathematical treatment of the experimental data has shown that such areas may be satisfactorily approximated by equation of the $R = kt$ type. Parameter k possesses the meaning of the colony growth rate and is used by the authors as the criterion of microorganism activity.

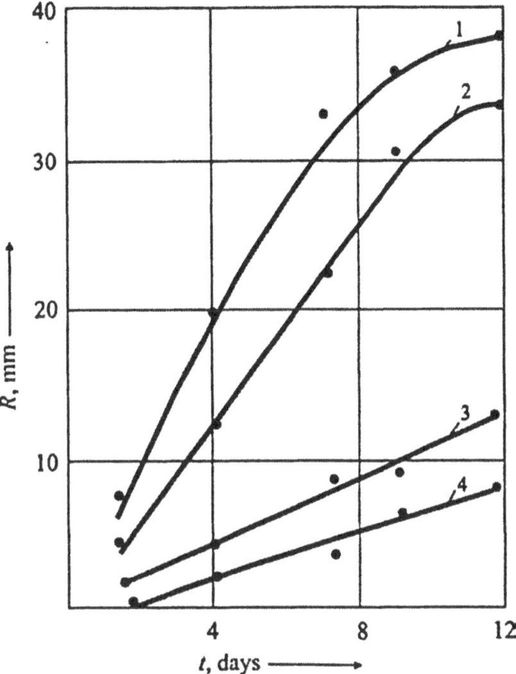

Figure A1-1. Dependence of the colony radius (R) of microscopic fungi on time of their cultivation in the Chapek-Dox medium at 29°C: 1 – *Aspergillus niger*; 2 - *Aspergillus amstelodami*; 3 – *Penicillium capsulatum*; 4 - *Penicillium cyaneo-fulvum*.

Figure A1-2 shows typical dependencies of k parameter on temperature of microscopic fungus cultivation.

The data obtained indicate that at corresponding temperatures the development intensity of all studied species of micromycete strains, separated under operation conditions, is higher than that of a collection culture. It has also been found that the growth rate of *Aspergillus niger* strain, detected on the wire insulation, exceeds development intensity of other studied species of fungi (by 3 – 4 times) in a broad range of temperatures.

The dependencies $k - T$ (Figure A1-2) are of the extreme type. Hence, the general feature of all tested strains of bioagents is the growth deceleration in the temperature range of 28 - 32°C. This temperature range is recommended for carrying out laboratory tests on fungal resistance of technical materials and articles. Optimal cultivation temperatures are shown in Table A1-2.

The data in Table A1-2 show that for many species of micromycetes, the optimal temperature ranges for growth of strains separated from machinery and collection ones are quite different. Obviously this is associated with adaptation of microorganisms to the real conditions of their dwelling on electric wire insulation in different regions.

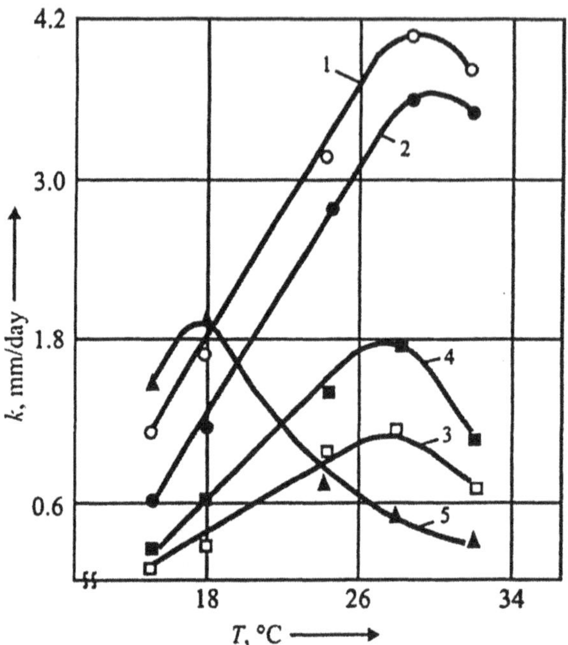

Figure A1-2. Dependence of the rate (*k*) of microscopic fungus colony growth in the Chapek-Dox medium on the cultivation temperature: 1 – *Aspergillus niger*; 2 - *Aspergillus amstelodami*; 3 – *Aspergillus niger* (collection strain); *Penicillium capsulatum*; 4 - *Penicillium cyaneo-fulvum*.

The results of studying characteristics of a microbiological factor allow the following recommendation to tests of the microbiological resistance of electric wire insulation: to use strains of microscopic fungi detected in suggested areas of the studied article operation as test-cultures. In this case, test modes should be selected with regard to the temperature range optimal for development of the current strain, as well as the temperature range, outside which the colony growth is partly or completely inhibited.

AGING OF POLYMERIC MATERIALS

The analysis of distribution of detected cases of microbiological damages with regard to the service time of electric wires induces a conclusion that the number of such damages of machinery under operation during 11 – 20 years is greater than that of articles operated during 1 – 10 or over 21 years. This induces a supposition about existence of an extreme dependence between microbiological resistance and the aging degree of materials used as insulation of electric wires.

The effect of PVC-plasticate and varnished fabric aging on *Aspergillus niger* growth on them has been studied. Samples of electric wires after different periods of operation (5 – 25 years) in OEC composition of technical articles have been studied. To reduce the limits of the experimental error, samples with unitypical construction location and operation conditions in the articles have been selected. In tests on PVC-plasticate, varnished fabric cover was removed from the samples. Fungi were cultivated on samples at 29°C and 100% humidity. The technique for determination of the biomass quantity on materials is indicated in Chapter 2 of the monograph.

Figure A1-3 shows kinetic curves of *Aspergillus niger* biomass growth on electric wire insulation of different operation periods (hereinafter, the period or aging duration).

The data shown in Figure A1-3 indicate that aging of PVC-plasticate and varnished fabric does not induce a change of the S-shaped type of $m - (t - L)$ curves. It is also obvious that the aging duration causes a change in slope of kinetic curves and maximal biomass (m_∞) reached during the fungus growth on the insulation. As a consequence, one may conclude that kinetic parameters of the microscopic fungus growth are sensitive to changes proceeding in materials during aging.

Mathematical treatment of obtained experimental data have shown that biomass growth on materials after different periods of aging as well as on initial (non-aged) samples is satisfactorily described by equation as follows:

$$m_t = \frac{m_\infty}{1 + a \cdot \exp[-b(t - L)]}, \tag{A1-1}$$

where m_t is the specific dry biomass at time t of microorganism cultivation on the material; m_∞ is the border (maximal) biomass (m_t) obtained in the experiment; L is the time, during which biomass remains constant on the material polluted with microbial cells, compared with its initial value; a and b are parameters of the biomass growth.

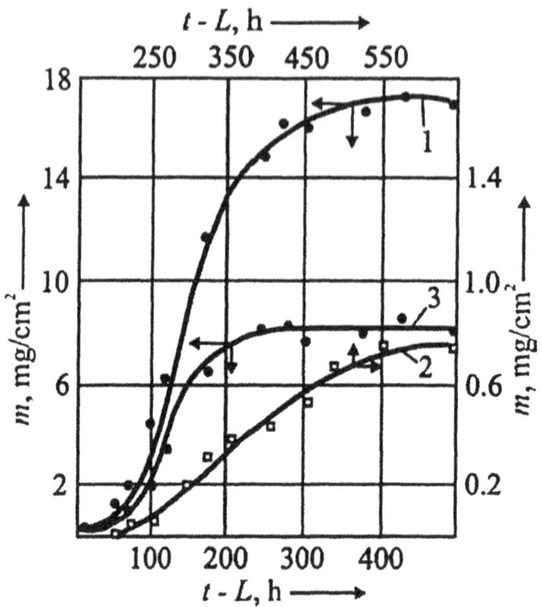

Figure A1-3. Kinetic curves of *Aspergillus niger* biomass growth on PVC-plasticate (□) and varnished fabric (•) of electric circuits of technical articles after 25 (1, 2) and 10 (3) years of operation.

Figure A1-4 shows dependencies of kinetic parameters of equation (A1-1) for *Aspergillus niger* biomass growth on electric insulation materials on the period of their aging. Clearly the values of a, m_0 and L of the process for the initial samples and the aged ones differ by 15 – 20%, i.e. the difference falls within the error range of their determination. As a consequence, the aging of studied materials is insignificant for values of a, m_0 and L.

As mentioned above, b and m_∞ parameters characterize bioresistance of the materials. They are sensitive to changes in PVC-plasticate and varnished fabric proceeding during aging (Figure A1-4) under operation conditions of electric wires.

Data shown in Figure A1-4 indicate that the specific rate of bioagent growth (b) and maximal biomass (m_∞) on the material increase monotonously with the period of varnished fabric aging. This means that the aging of varnished fabric reduces its microbiological resistance.

On the contrary, the values of b and m_∞ parameters for PVC-plasticate, the operation period of which is less than 5 years, are close to those obtained for micromycete growth on the initial (non-aged) material (Figure A1-4). During this period, an increase of sample aging duration proceeds with some increase of the parameters considered. This coincides with the above-mentioned supposition about an extreme dependence between the number of cases with biodamages and the operation period of articles.

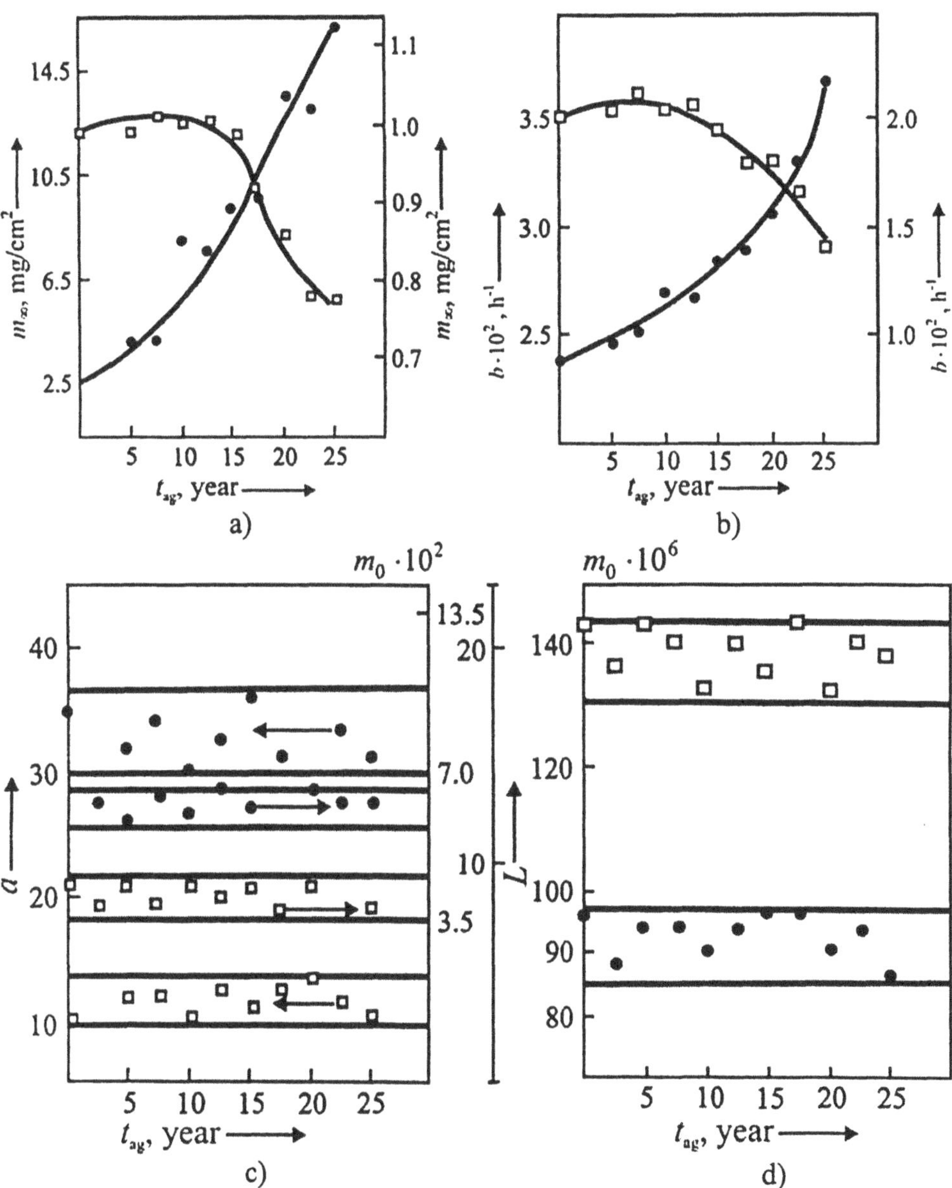

Figure A1-4. Dependence of parameters m_∞ (a), b (b) a, m_0 (c) and L (d) on the period of aging (t_{ag})of PVC-plasticate (□) and varnished fabric (•).

Insulation samples from PVC-plasticate operated in articles over 5 years are characterized by a reduction of intensity of *Aspergillus niger* growth on them. For such samples, the values of b and m_∞ parameters are smaller than in case of the fungus growth on non-aged plasticate. As a consequence, if the aging period of PVC exceeds 10 years, microbiological resistance of it is increased.

Obviously, differences detected in the type of aging effect on the resistance of insulation materials to bio-overgrowth are associated with specificity of chemical and physical processes proceeding in PVC-plasticate and varnished fabric. One may suggest that after 10 years of aging, reduction of interaction intensity between biodestructor and insulation from polyvinylchloride is first stipulated by reduction of the plasticizer content in PVC-plasticate, which is the main source of hydrocarbon for microorganisms.

The effect of environmental factors on varnished fabric induces thermooxidative degradation of nitrocellulose cover. As a result, low-molecular fragments of macromolecules are formed, which are easily assimilated by microorganisms as the nutrition source. Concentration of these fragments in the material is increased with the period of aging, i.e. substances nutritive for bioagent are accumulated. This, in its turn, reduces resistance of varnished fabric and induces microbiological overgrowth.

The results obtained during studying the effect of aging (Figures A1-3 and A1-4) and temperature-humidity conditions of the bioagent cultivation (Chapters 4 and 5 of the monograph) on bio-overgrowth of electric insulation materials have been compared. It has been found that the values of b and m_∞ parameter variations obtained for microorganism development on aged samples in relation to initial (non-aged) ones and applying different temperature-humidity test modes are comparable. For the studies ranges of variations of the mentioned parameters, the values of b and m_∞ parameters may be reduced or increased by 2 – 6 times. This confirms a conclusion obtained with the help of a priori method of analysis that besides the temperature-humidity complex of the environment, aging of electric insulation induces a significant effect on development of its microbiological damaging.

The dependencies shown in Figure A1-4 allow determination of kinetic parameters of equation (A1-1) for biodestructor growth on the studied insulation materials after different periods (within the studied range) of their aging, i.e. biological resistance of operated electric insulation.

Such estimation may be performed graphically according to dependencies shown in Figure A1-4, as well as with the help of the correlation equations associating b and m_∞ parameters with time of the material aging (Table A1-2).

Mathematical treatment of experimental data shown in Figure A1-4 makes possible determination of aged insulation capability of inducing growth of microscopic fungi under conditions of 100% air humidity and 29°C, i.e. under test conditions provided by scientific and technical data on estimation of biological resistance of technical materials and articles active at present.

Table A1-3

Correlation equations describing dependencies of b and m_∞ parameters on duration of electric insulation material aging (aging period up to 25 years)

Material	Type of equation	Values of coefficients
Varnished fabric	$b^{ag} = b^0 \cdot \exp(k_b \cdot t_{ag})$	$k_b = 1.2 \cdot 10^{-2}$ year^{-1}
	$m_\infty^{ag} = m_\infty^0 \cdot \exp(k_m \cdot t_{ag})$	$k_m = 1.0 \cdot 10^{-1}$ year^{-1}
PVC-plasticate	$b^{ag} = b^0$	
	$m_\infty^{ag} = m_\infty^0$	
	$b^{ag} = b^0 - k_b(t_{ag} - L')$	$k_b = 3.0 \cdot 10^{-4}$ year^{-1}
	$m_\infty^{ag} = m_\infty^0 - k_\infty(t_{ag} - L')$ for $t_{ag} > 5$	$L' = 5$ years

Note: b^{ag}, m_∞^{ag}, b^0, and m_∞^0 are parameters of equation (A1-1) for *Aspergillus niger* biomass growth on preliminarily aged during time t_{ag} and initial (non-aged) electric insulation materials at 29°C and 100% humidity.

To calculate parameters of the fungus growth on aged insulation under other corresponded to real conditions of OEC temperature-humidity operation modes, besides those shown in Table A1-2, equations presented in Chapter 5 of the monograph may also be used.

Note that empirical correlation expressions and values of their coefficients and parameters may be applied to description of the bioagent development only on studied types (trademarks) of PVC-plasticate and varnished fabric.

POLLUTION COMPOSITION

The following variants of the interaction between microorganisms-destructors and materials met in real operation have been modeled:

1. in the absence of pollution. For this purpose, samples have been inoculated by a suspension of fungus spores in distilled water (mode 1);

2. in the presence of mineral pollution. Suspension of spores in the Chapek-Dox medium without saccharose has been used for pollution (mode 2);
3. in the presence of mineral and traces of organic pollutions. Samples were polluted with a suspension of fungus spores in the Chapek-Dow medium containing 0.05% saccharose (mode 3);
4. in the presence of mineral and organic pollutions. Samples were polluted by a suspension of spores in the Chapek-Dox medium (mode 4).

Test results (Table A1-4) indicate that growth of microorganisms is observed on all studied materials both in the presence and in the absence of pollutions, i.e. micromycetes are capable of using these materials as nutritious sources.

Table A1-4

Estimation results of *Aspergillus niger* impact on materials in the presence of various **compositions of pollutions (test duration is 180 days)**

Material	**Micromycete development degree, deg. according to GOST 9.048 (P/P_{cont})**			
	Mode 1 (no pollution)	Mode 2 (mineral pollution)	Mode 3 (mineral and trace organic pollution)	Mode 4 (mineral and organic pollution)
Fiberglass laminate	3 ($1.2 \cdot 10^{-1}$)	4 ($1.6 \cdot 10^{-1}$)	5 ($2.1 \cdot 10^{-1}$)	5 ($1.0 \cdot 10^{-1}$)
Varnished fabric	4 (0.3)	5 (0.3)	5 (0.2)	5 (0.2)
PVC-plasticate	2 (1.6)	2 (1.8)	3 (2.1)	3 (2.4)
Cotton thread	5, thread destruction	—	—	—

Note: — - no tests have been performed.

For all tested materials, the presence and composition of pollutions induce a significant impact on efficiency of the microbiological damaging.

The presence of a source of carbon and mineral components easily assimilated by microorganisms (modes 3 and 4) on PVC-plasticate, varnished fabric and fiberglass laminate promotes more intensive overgrowth of these materials. Hence, the data from Table A1-4 indicate that the microorganism growth degree on these materials under modes 3 and 4 is practically the same. As a consequence, the process intensity is independent of the quantity of carbon source in the pollution, easily assimilated by micromycetes.

If saccharose is completely absent in the model pollution, total quantity of biomass on samples is reduced (mode 2). In this case, the material itself represents the source of carbonic nutrition of microorganisms. At the same time, the effect of micromycetes on electric properties is slightly different in tests in modes 2, 3 and 4. Obviously, the components easily assimilated by microorganisms stimulate growth of fungi, which do not change the type of processes deteriorating properties of the materials. This fact is also proved by comparison of the tests results obtained for modes 1 and 2. Micromycete growth under the mode 1 confirms ability of the material to provide vital activities of fungi. The presence of pollutions (mode 2) stimulates these processes. Hence, variations of electrical resistance (*R*) under modes 1 and 2 differ insignificantly.

THE EFFECT OF TECHNOGENIC MAGNETIC FIELDS

The measurement of magnetic field intensities on 8 types of articles have shown that there is a magnetic field around studied objects, which is 10 – 80 T intensive and, on average, exceeds the geomagnetic field in the present locality by 10 – 30%.

Magnetometric changes inside the articles have indicated that the field intensity is significantly different along their longitudinal axis. Hence, all articles could be conditionally separated into several analogous zones with typical and comparatively constant intensity of magnetic field.

Revelation of microbiological damages on materials and articles in every set zone indicated that the quantity of biodamages increased with the magnetic field intensity within its technogenic range (50 – 80 T).

Note that the occurrence of biodamages under real operation conditions of the machinery may be stipulated by a totality of a series of arbitrary reasons: local increase of air humidity and/or temperature, occurrence of various pollutants on the surface of the article, disturbance of the ventilation mode, etc. As a rule, there is no information about prehistory on occurrence of each case of biodamages. In this connection, statistic data obtained during investigations of technical conditions of the articles may not yet be enough proof of the stable dependence between intensities of real technogenic magnetic fields and biological damaging of the articles.

The presence of such dependence has been proved by laboratory tests of the effect of magnetic field intensity on *Aspergillus niger* development.

Industrial permanent magnets shaped as Helmholtz rings were used as the sources of magnetic fields. Field intensities from 20 to 400 T were used in the experiments. Magnetic field intensities were measured by *MERTA* magnetometer.

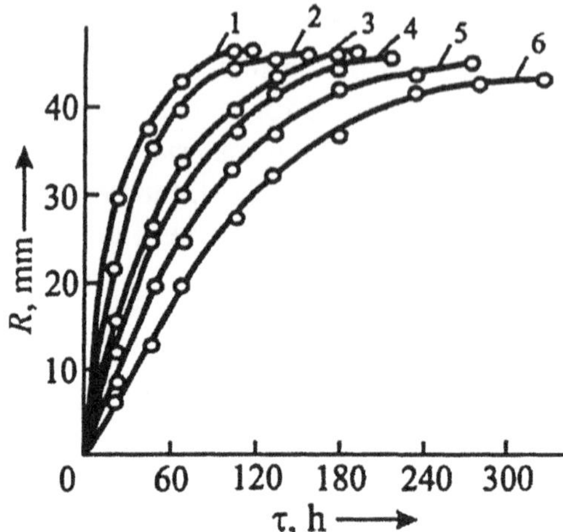

Figure A1-5. Kinetic dependencies of *Aspergillus niger* colony growth on nutritious media: 1, 3 – wort-agar; 2, 5 – Chapek-Dox medium; 4, 6 – meat – peptonic agar; 3, 5, 6 – without the effect of magnetic field; 1, 2, 4 – with 80 T magnetic field effect.

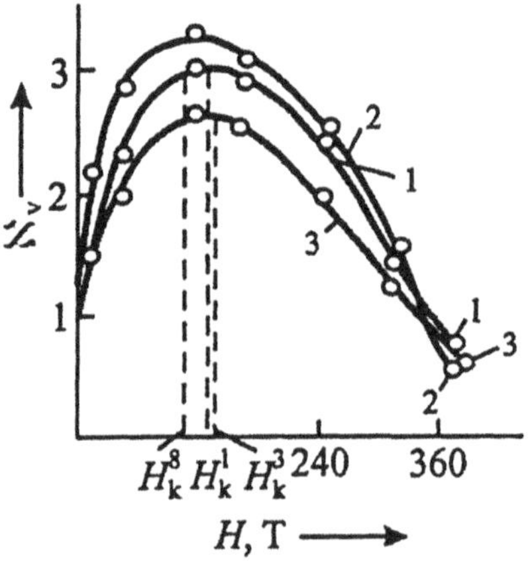

Figure A1-6. Dependence of criterion of the magnetic field effect efficiency (ℵ) during *Aspergillus niger* development on nutritious media on magnetic field intensity (80 T): 1 – Chapek-Dox medium; 2 – meat – peptonic agar; 3 – wort-agar.

Figure A1-5 shows several kinetic dependencies of *Aspergillus niger* colony growth during its development on different nutritious media under the effect of magnetic field.

Based on displayed calculations, all presented dependencies are successfully described by the equation of the following type (Emanuel, 1977):

$$R = R_{\infty}[1 - \exp(-K\tau)], \tag{A1-2}$$

where R is the radius of the colony at time, τ; R_{∞} is the border value of the colony radius, reached under current conditions of the experiment; K is the rate constant of the colony growth; τ is the time of the colony growth ($\tau = t - L$, where t is the time since nutritious medium sowing; L is the lag-phase duration, i.e. the period, during which no growth of the colony is observed. In all experimental data displayed, the lag-phase duration is 24 hours).

For one and the same microorganism-nutritious medium system, the K constant depends on the field intensity (H) only and, as a consequence, characterizes its effect on the rate of microbial cell growth. The efficiency of this effect was estimated by the dimensionless parameter as follows:

$$\aleph = K_H/K_0, \tag{A1-3}$$

where K_0 is the rate constant of the colony growth without effect of magnetic field; K_H is the rate constant of the colony growth in magnetic field with intensity H.

The criterion of the field effect efficiency, $\aleph_v$, shows by how many times the fungus growth is accelerated (or decelerated) under the effect of magnetic field.

The rate constant (K_0) of *Aspergillus niger* colony growth on Chapek-Dox, MPA and WA media, calculated by equation (A1-2), equaled $1.3 \cdot 10^{-2}$, $0.7 \cdot 10^{-2}$, and $2.0 \cdot 10^{-2}$, respectively. The dependence of the $\aleph_v$ criterion on magnetic field intensity on the same nutritious media is shown in Figure A1-6.

Figure A1-6 indicates that graphical dependencies of $\aleph_v$ on the magnetic field intensity are of the extreme type. As a consequence, there is a definite value of magnetic field (H_k), at which acceleration of the growth rate of *Aspergillus niger* is maximal.

The use of different nutritious media induces somewhat variation of magnetic field intensity (H_k, Figure A1-6), which equals 120 T (H_k) for the Chapek-Dox medium, 150 T (H_k^3) for the wort-agar, and 100 T (H_k^2) for the meat-peptonic agar. On the average, the growth rate of *Aspergillus niger* colony under the effect of these magnetic field intensities was increased by 2.6 – 3.4 times.

Of special interest is the fact that at constant intensity of magnetic field ($H \leq 320$ T) $\aleph_v$ depends on the type of the medium used (Figure A1-6). Hence, maximal $\aleph_v$ value was observed for *Aspergillus niger* development on the meat-peptonic agar medium. It is somewhat lower for the Chapek-Dox medium, and reaches its minimum on the wort-agar. At the same time, K_0 was maximal for the wort-agar and minimal for the meat-peptonic agar.

Hence, data obtained indicate the presence of a regularity, according to which an increase of microscopic fungus growth rate induced by technogenic magnetic fields may be so higher, the lower is favorability of the substrate used for its growth.

Compared with the control value $\aleph_v < 1$ (Figure A1-6), the growth rate of *Aspergillus niger* was reduced, when magnetic field intensity exceeded 350 T. As a consequence, constant magnetic fields over 350 T intensity may suppress growth of some species of microorganisms.

The results shown in Figure A1-6 also indicate that the effect of magnetic field of 20 – 80 T intensity has induced a significant (1.5 – 3-fold) increase of *Aspergillus niger* growth intensity on nutritious media. This very field intensity was observed in the areas of studied articles, mostly subject to biodamages. Hence, constant magnetic fields with intensities close to the real ones affecting the articles may accelerate development of micromycetes, which, in its turn, may be the reason of biodamaging intensification of materials and articles from them.

Taking into account that intensity of the fungus growth on materials used in machinery is much lower than on nutritious media (Sadauskas *et al.*, 1987), one may suggest that magnetic field may cause a significant effect on bio-overgrowth of technical materials.

According to GOST 9.048-75, visual estimation of overgrowth of tested linen and cotton threads has shown that development of *Aspergillus niger* impacted by technogenic magnetic field and in the absence of it on both types of threads equals 5 degrees. However, the highest intensity of the fungus development on linen threads has been observed in case of the magnetic field effect (H = 80 T). Visual inspection of linen threads indicated over 70% overgrowth of their surface, whereas cotton threads displayed less 50% level. More intensive overgrowth of linen threads compared with cotton ones is apparently associated with more dense weaving of the latter that reduces the surface of their contact with developing micromycetes.

The effect of technogenic magnetic fields has also increased aggressiveness of the microscopic fungus to the materials, i.e. the effect of biofactor on properties of the studied thread. The data shown in Figure A1-7 indicate that reduction of strength at break (σ) of linen and cotton threads during *Aspergillus niger* development in magnetic field (40 – 80 T) proceeds 1.5 – 2 times faster than in case of the field absence.

The data shown in Figure A1-7 were analyzed by their approximation by equation (A1-4) applied in the reference (Moiseev, Zaikov, 1979) to description of polymer biodegradation as follows:

$$\sigma = \sigma_0 \left(1 - \frac{K_{eff}}{r\rho}\tau\right)^2, \tag{A1-4}$$

where σ_0 is the initial strength at break of the thread; τ is time of the colony development on the material ($\tau = t - L$, where t is the time passed since inoculation of the material; L is the lag-phase duration); r is the thread radius; ρ is the thread density; K_{eff} is the constant characterizing the rate of thread strength variation induced by microorganisms.

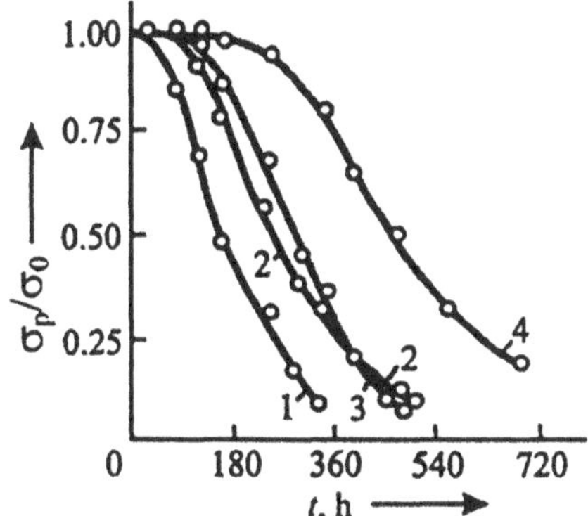

Figure A1-7. Dependence of relative variations of linen (1, 2) and cotton (3, 4) thread strength on time of *Aspergillus niger* development on them: 1, 3 – in constant magnetic field (80 T); 2, 4 – without field.

In this case, the dimensionless criterion $\aleph$ may be presented as follows:

$$\aleph = \frac{K_{eff}^{mf}}{K_{eff}^{0}}, \tag{A1-5}$$

where K_{eff}^{mf} and K_{eff}^{0} are constants characterizing the rate of thread strength variation in magnetic field and in the absence of it, respectively.

Table A1-5 shows values of K constant and $\aleph$ criterion calculated by equation (A1-5).

Table A1-5

K_{eff} constant and ℵ criterion values for strength variations of threads damaged by *Aspergillus niger*

Determined parameters	Thread type	
	Linen	Cotton
K^{0}_{eff}, h^{-1}	$0.40 \cdot 10^{-6}$	$0.22 \cdot 10^{-6}$
K^{mf}_{eff}, h^{-1}	$1.14 \cdot 10^{-6}$	$0.77 \cdot 10^{-6}$
ℵ	2.85	3.50

In the absence of magnetic field, the rate constant of cotton thread strength variation was found equal approximately 1.5 times lower of the constant corresponded to linen threads. Obviously, this is associated with the specificity of their texture (weaving) that was mentioned above.

Values of ℵ criterion (efficiency of the fungus growth on threads) are approximately corresponded to $ℵ_v$ values obtained for *Aspergillus niger* growth on microbiological nutritious media. Hence, it should be noted that magnetic field has mostly accelerated biodamaging of cotton threads, which more strong and resistant to the effect of fungi. This corresponds to the detected effect of higher acceleration of *Aspergillus niger* colony growth on media less favorable for its development under the conditions of technogenic magnetic field effect. That is why one may suggest that the increase of fungus aggressiveness to materials studied in magnetic field is associated with the acceleration of its growth.

Hence, the investigations carried out indicate that technogenic magnetic fields may cause an effect on biodegradability of technical materials by accelerating the growth rate and aggressiveness of *Aspergillus niger*. There is a correspondence between distribution of magnetic field intensity in different parts of articles and the quantity of biodamages in them. The results obtained should be considered in development of laboratory methods of estimation of biological resistance of materials and articles from them and machinery maintenance under real operation conditions. Note also that constant magnetic field (over 350 T intensity) may decelerate growth of microscopic fungi. This effect used in development of the measures preventing and eliminating biodamages of the machinery.

Appendix 2.

Kinetics of radial growth of Aspergillus colonies at different temperatures

Colonies of three Aspergillus species reached the highest radial growth rate at 30 - 40°C and these rates were constant during some days. At 20 and 45°C the radial rates were smaller than at 30 - 40°C in spite of their increase during 8 - 10 days of colonies growth. The accelerated radial growth at 20 and 45°C may be explained by adaptation of the fungi to unfavorable temperatures. The calculations show that at the constant growth rate the surface area occupied by colonies is increased with the constant acceleration and with its increase – with the growing acceleration. It is supposed that the process of the mycelial micromycete affection of different materials which proceeds by the type of formation of isolated colonies on their surface tends to the acceleration development at the constant temperature.

Many authors [374, 376, 379, 380, 386] have shown that the radial rate remains constant during the greater part of the radial growth of mycelial micromycete colonies on the dense nutritious media surface. It is considered that such uniform radial growth proceeds until the colony border reaches the nutritious medium border (the "linear growth law"). However, in these investigations fungi grew at close to optimal temperatures, whereas the radial rate, average for the total period of the radial growth, mostly depends on temperature [377].

Aspergillus flavus Link ex Fr., *Aspergillus niger* Thiegh and *Aspergillus terreus* Thom fungi strains were received from the collection of the Institute of Microbiology, AS BelSSR. Agarized Chapek medium [375] was poured out in portions (20 ml) to Petri dishes with 9 cm internal diameter. The dishes were inoculated in the center of the medium surface by a suspension of fungus conidia with further incubation at 10, 20, 30, 40, 45 and 50°C in Ts-1241-MU 4,2 incubators, temperature in which may be kept at a level below the environmental one. Since the beginning of the incubation, diameters of five colonies in every variant of the experiment were measured every 24 hours in perpendicular directions. During measurements, the dishes were kept at 22 - 25°C less than 30 minutes. The data obtained were statistically treated by Ashmarin and Vorobiev [373], and differentiated by Tsyipkin [383]. The capability of Aspergillus of utilizing organic acids as the unique source of carbon and energy was determined by sowing them to test-tubes with liquid Chapek medium, to which 1 wt.-vol.% citric, acetic or malic acid ammonium salt was injected instead of carbohydrate.

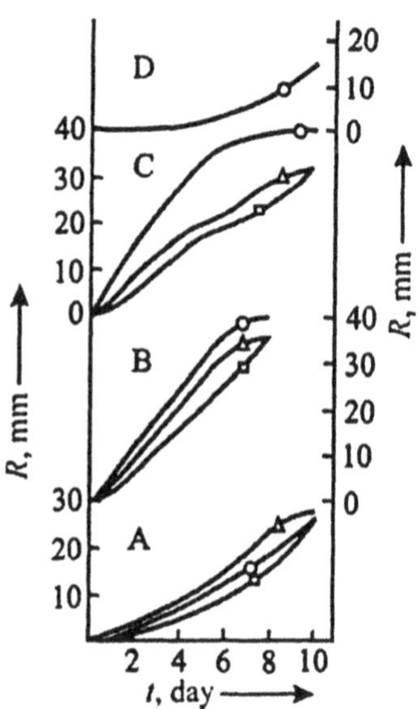

Figure A2-1. Dependence of colony radius of Aspergillus (R) on growth duration (t): A – at 20°C; B – at 30°C; C – at 40°C; D – at 45°C; Δ – *Aspergillus flavus*; ○ – *Aspergillus niger*; □ – *Aspergillus terreus*.

At 20 - 45°C *Aspergillus niger* formed colonies and the rest two species of Aspergillus at 20 - 40°C only. The graphs characterizing dependence of the average radius length (R) on the colony growth duration (t) in all 10 variants of the experiment, in which fungi have formed colonies, are shown in Figure A2-1. At the specified probability of 0.95 after 3 – 4 day of the dish incubation, the confidence range of the average radius length (hereinafter, the radius length) falls within ±5%. In Figure A2-1 scales, it is impossible to display so low variability. The Figure shows that the greater part of the graphs characterizing the colony growth at 30°C represents straight lines. As a consequence, during the greater part of 8 days of the colonies observation at 30°C their radii were increasing at a constant rate. The uniform radial growth of *Aspergillus terreus* colonies lasted up to 8 days, reaching 37.5 ± 2.5 cm during this period. The radii of the colonies of the rest two species reached this value already after 6 days, after which the radial growth rate decelerated. Similar to deceleration of the microbial culture growth after the exponential stage [380, 381], this deceleration of the radial growth rate may be explained by the reduction of nutritious substance concentration in the medium and accumulation of fungus metabolites in it.

The radius of *Aspergillus niger* colonies at 40°C increased uniformly up to the fourth day only, the radial growth rate being somewhat higher in this period compared with 30°C. At 40°C the radial growth rates of *Aspergillus flavus* and *Aspergillus terreus* colonies between the fourth and the sixth days were temporarily decelerated. This may be explained by the reasons analogous to those inducing variations in the rate of culture development of various microorganisms in liquid nutritious media [380, 381, 384]. Many species of mycelial micromycetes are capable of forming organic acids as the products of incomplete oxidation of carbohydrates [384], and studied species of Aspergillus genus are capable of assimilating citric, acetic and malic acids are the unique source of energy and carbon. These fungi may probably produce these acids, which, in its turn, reduces pH of the medium and decelerates the radial growth of the colonies. Thereafter, when glucose concentration in the medium was reduced, the fungi began complete oxidation of carbon and acids, which increased pH of the medium and corresponded acceleration of the radial growth rate. The mentioned variations of the radial rate are low and short-term so that the general type of the radial growth in these two variants is closer to linear than to the accelerated one. For 10 days at 40°C the average radial growth rate of *Aspergillus flavus* and *Aspergillus terreus* colonies was lower than that for 8 days at 30°C.

At 45°C *Aspergillus niger* visible colonies occurred at the 3rd or 4th day of the dish incubation. The graph characterizing the radial growth in this variant represents a curve, the slope of which to the abscissa axis increases with time. As a consequence, since the beginning of growth till the 10th day the radial growth rate increased, but did not reach values, obtained at 30 or 40°C. At 20°C the radial growth rates of all three Aspergillus species also increased up to 8-10th day, but did not reach the values, obtained at 30 - 40°C. Thus among all tested temperatures, the optimal ones for the radial growth of *Aspergillus niger* colonies is 40°C and 30°C for the colonies of the rest two fungi. Hence, for these fungi, 30 and 40°C are the more favorable temperatures than 20 and 45°C.

The radial growth of mycelial micromycete colonies proceeds as a result of elongation of the hypha located on the periphery of their radii, which is stipulated by growth and end cell division [379, 387]. Obviously at favorable temperatures the end cells of peripheral hypha grow and divide at a constant rate, maximum possible under current conditions until changes in the composition of nutritious medium begin decelerating this process. That is why at such temperatures the radial rate of the colony growth is mostly constant. At unfavorable temperatures these cells grow and divide slower than at favorable ones, but the rate of this process may be somewhat increased as a result of adaptation. At 20 - 45°C the surface area of the nutritious medium occupied by the fungus colonies (hereinafter, the area), preserved the shape of more or less regular circle. This means that the end cell growth and division rates of all peripheral hyphas of the colony are equal and all cells adapt to unfavorable temperatures simultaneously.

The regular radial growth of the colonies is characterized by the known equation (A2-1) [378, 380, 387] as follows:

$$V_r = dR/dt = (R - R_0)/(t - t_0) = \text{const}, \qquad \text{(A2-1)}$$

where V_r is the radial rate; R and R_0 are the colony radii at moments t and t_0 since the growth initiation, respectively.

Using this equation and sequentially differentiating the colony area ($G = \pi R^2$) and the growth rate of it (V_g) by time, one may display that $V_g = 2\pi R V_r$ and at the constant radial rate $A_g = 2\pi(V_r)^2$, where A_g is the acceleration of the colony growth rate. As a consequence, at the constant radial rate the area of the colonies grows at the constant acceleration, and acceleration is increased with the radial rate.

Among other factors, the colony biomass depends on its square, and the quantity of metabolites formed on its biomass [380, 385]. Obviously at the constant radial rate the colony biomass and the quantity of metabolites formed will tend to accelerated growth. This fact should be taken into account in the study of kinetics of the mycelial micromycete colony biomass growth and formation of metabolites such as enzymes and organic acids, which are considered the main factors of mentioned fungi damaging industrial materials. In some cases, molding of foods and industrial materials proceeds similar to formation of isolated colonies on their surface [382], which is clearly seen on photos [372]. One may suppose that at the constant temperature during greater part of the molding development of this type the total area of the material surface and quantity of metabolites formed will tend to accelerated growth. As follows from these data, molding may be decelerated at high temperature variations and other unfavorable changes in the environment.

Appendix 3.

Kinetics of acids excretion in medium by mycelial micromycetes colonies

---

The data on kinetics of the mycelial micromycete colonies growth on the surface of dense nutritious media are shown in many publications [137, 391, 396, 390, 393 – 395, 401 – 403]. It is also shown that the growth of *Aspergillus penicilloides* surface colonies is accompanied by the acids excretion to the medium [398]. It seemed significant to study acids excretion by colonies of different species of micromycetes.

Applied in experiments were 9 strains of the following fungal species from the collection of the Institute of Microbiology, AS BelSSR: *Aspergillus flavus* Link ex Fr., *Aspergillus niger* van Thieghem, *Aspergillus terreus* Thom, *Chaetomium globosum* Kunre ex Fr., *Paecilomyces varioti* Bain, *Penicillium chrysogenum* Thom, *Penicillium verrucosum* Dierkx *var. cyclopium* Westlins, *Penicillium funiculosum* Thom, *Trichderma viride* Pers ex S.F.Grey [= *Trichoderma lignorum* (Tode) Harz].

A medium of the following composition (g/l) was prepared: $NH_4H_2PO_4$ – 0.1; KCl – 0.2; $MgSO_4{\cdot}7H_2O$ – 0.02; agar-agar – 0.15. Bromcresol purple indicator up to 0.008 g/l and after sterilization – sterile solution of saccharose up to 0.05 g/l concentration were added to the medium [389]. It is known that alkaline solutions of the current indicator are of purple-violet color, and the acidic one is yellow; hence, color is changed within pH range of 6.0 – 6.5 [392]. Melted medium was poured out in portions (20 ml) to Petri dishes of 9 cm internal diameter. When the medium was cooled down, the dishes were inoculated by fungal conidia (5 dishes per each species). Conidia were injected to the center of the medium surface, after which the dishes were incubated during 10 days at 30°C [395].

Figure A3-1 shows that the total duration of the lag-phase and sequential phase of accelerating radial growth of the colonies of all 9 fungal species equaled several ten hours, but no longer than a day in the majority of cases. After that the colonies reached the maximal radial growth rate, which remained constant till the end of incubation or till the colony border reached the border of the nutritious medium. The areas of the graphs characterizing this phase of growth are linear. The variation coefficient of the average radius length of 5 analogous colonies did not exceed 5% at any moment of measurements. So low variability did not allow graphical presentation of the squared error. The radii of *Aspergillus terreus*, *Penicillium chrysogenum* and *Trichoderma viride* colonies increased approximately 1.5 times slower than before on the Chapek-Dox-agar medium [393]. For the rest fungi the difference in the radial rates on these media is smaller or practically absent. Slower growth of some fungi in this experiment may be explained by the presence of indicator (5,5-dibrom-o-cresol sulfophathalein) in the medium [106] and 6-fold lower carbohydrate concentration. It is known that chemical compounds containing phenolic groups and halogens possess anti-fungal activity, which is so higher, the lower carbohydrate concentration is [394].

Hence, during the greater part of the incubation period the colonies grew in accordance with the "law of linear growth" [137, 402, 403]. Kinetics of linear, i.e. regular growth may be described by the equation as follows:

$$V_r = \mathrm{d}R/\mathrm{d}t = \text{const}, \qquad \text{(A3-1)}$$

where V_r is the radial rate; R is the colony radius; t is the period (time) since the incubation beginning. Such kinetics is typical of mycelial micromycetes colonies growing on usual agarized nutritious media under optimal or almost optimal conditions. The radial rate is often denoted as K_r meaning its constancy [137]. Taking into account that it is not always constant [395, 395], it should be preferably denoted as V with an appropriate index, as it is usual for the rates of various processes [280].

The radial growth of mycelial micromycete colonies is caused by hypha elongation, located radially on the colony periphery; the hypha elongation is caused by growth and fusion of the end cells [396, 403]. Inoculate conidia germinate during two initial phases of the radial growth forming hyphas, which, in their turn, branch and elongate forming a colony. Hence, the growth and fusion rate of the end cells adapting under the environmental conditions is increased from zero to some maximum.

During linear growth, the growth and fusion rate of peripheral hypha end cells and, consequently, the radial growth rate of the whole colony remain constant. Therewith, the colony profile parallel to the dense nutritious medium, which is often incorrectly named the "square" of it [137, 404], usually preserve the circular shape. As a consequence, momentary growth and fusion rates of the end cells of all peripheral hyphas are the same. Differentiating the colony square (G_c) and the rate of its increase (V_{gc}) by time and substituting the radial rate value to equation (A3-1), one may display that

$$V_{gc} = 2\pi R V_r, \tag{A3-2}$$

and at constant radial rate

$$A_{gc} = \pi(V_r)^2 = \text{const}, \tag{A3-3}$$

where A_{gc} is the process acceleration [394]. Total duration of the initial two phases of the radial growth is approximately 10 times shorter under favorable conditions than that of the linear growth. To simplify kinetic calculations, non-linearity of the initial two phases may be neglected and the radial rate may be assumed constant since the very beginning of the colonies growth. Then equation (A3-3) is easily transformed as follows:

$$G_c = A_{gc}t^2/2 = k_1t^2, \tag{A3-4}$$

where $k_1 = A_g/2 = \text{const}$.

Equations (A3-2) – (A3-4) indicate that during the linear radial growth the colony square is increased at the constant acceleration, and momentary value of the square is proportional or approximately proportional to square time.

On usual dense nutritious media, the growth of mycelial micromycete surface colonies and the majority of other microorganisms perpendicular to the medium surface is limited. That is why such colonies are considered in kinetic calculations as a flat cylinder, the height (h) and the average physical density (ρ) of which are constant, and the volume and the mass increase with the basement square [137, 404]. As a consequence, kinetics of the colony volume and the mass (M_c) growth under such conditions will be analogous or very close to kinetics of its square increase. Hence, in particular,

$$M_c = h\rho G_c = k_2t^2, \tag{A3-5}$$

where $k_2 = k_1h\rho = \text{const}$. Based on the experimental data [401], the calculations have been performed indicating the mass increase of such colonies according to equation (A3-5).

Immediately after the occurrence of visible colonies on the medium surface, its color beneath and around them was changed from purple-violet to yellow. As a consequence, all 9 fungi excreted acids (or acid) to the medium, which happened simultaneously with the occurrence of visible colonies. As the medium is uniformly thick, the zones of the color change represented flat cylinders, on the upper basement of which fungal colonies were located concentrically. Variability of the average radius length of 5 analogous zones at any moment did not exceed 0.5%, and the radii of medium color change zones were always greater than those of the appropriate colonies (Figure A3-2). The radii of the mentioned zones increased uniformly during linear radial growth of the colonies. Thus, kinetics of the increase of the basement area and the color change zone volume of the medium was analogous to that of the colony mass and square increase.

Motion of the color change zone border characterizes diffusion of acids to the nutritious medium periphery. In accordance with the first Fick law, the diffusion rate of substance is proportional to its concentration gradient [280]. As a consequence, at the constant radial rate of these zones increase, the acid concentration gradient at their borders was also constant, and the acid concentration increase in the medium was compensated by the zone volume growth. This means that the quantity of acids in the color change zones of the medium increased in proportion to increasing volume of the zones and mass of the colonies (see equation (A3-5)), i.e.

$$P_c = k_3 M_c = k_4 t^2, \qquad \text{(A3-6)}$$

where P_c is the quantity of acid in the color change zone of the medium; k_3 = const is an empirical coefficient; $k_4 = k_2 k_3$ = const.

However, theoretical model of the microbial biosynthesis product excretion is known, based on a logical suggestion that the quantity of cells excreting the product is increased proportionally with the microorganism-producer mass. This model is characterized by the equation as follows:

$$V_p = \mathrm{d}P/\mathrm{d}t = qM, \qquad \text{(A3-7)}$$

where V_p is the product excretion gross rate and q is the product excretion specific rate, acids, for example [137, 239]. Substituting the mass from equation (A3-5) to equation (A3-7) and integrating the expression obtained, one may check the validity of this model in relation to mycelial micromycete colonies. As acids were absent in the medium before initiation of the colony growth and the integration constant equals zero, integration gives the equation as follows:

$$P_c = qk_2 t^3/3 = k_5 t^3, \qquad \text{(A3-8)}$$

where P_c is the acid quantity; $k_5 = qk_2$ = const. Equation (A3-8) indicates that if all cells of the colony would excrete acids, its quantity in the medium would increase with cubic time. This would increase the acid concentration in the color change zone of the medium, the concentration gradient at the border of the zone and accelerate the increase of its radius. Hence, the graphs representing dependence of the color change zone radii on time would be quadratic parabola shaped (the curves with increasing slope to the abscissa axis) but not strait lines. As a consequence, accumulation of acids in the medium proceeded slower than it was expected from the theoretical model. The only thing is to suggest in this case that not all cells of the colonies excrete acids.

Biosynthesis of citric, malic and other acids by fungi is tightly associated with the Crebbs cycle [132]. As a consequence, these acids are synthesized by actively metabolizing fungal cells, and the rate of acid biosynthesis is decelerated with their vital activity. According to morphophysiological signs, the mycelial micromycete surface colonies formed are subdivided in, at least, two parts. The central part is formed by completely developed mycelium, including conidia formation. This part of the colony is surrounded by the growth zone circle, where active hypha branching and elongation proceed [137, 396, 402, 403]. It is known that 97% of ^{32}P-phosphate present in the medium is absorbed by cells from the growth zone and only 3% by cells of the central zone [401]. ^{14}C-carbohydrate labeled glucose and N-acetyl glucosamine are mostly absorbed by the end cells of growing hyphas [400]. All these facts allow a supposition that the acids are excreted only or mostly by actively metabolizing fungal cells accumulated in the peripheral zone of the colony growth.

To check this supposition, the surface of the Zvyagintsev medium in dishes was covered by cellophane discs, on which fungal colonies were grown in accordance with the above-mentioned method. The medium components and organic acids excreted by the fungi represent low-molecular substances and diffuse through the cellophane discs. That is why the colonies grew on cellophane, and the medium beneath it changed color.

Generally, kinetics of the colonies growth and the zone increase was analogous to kinetics of these processes in dishes without cellophane. When the colonies radii reached 3 – 3.5 cm, cellophane discs with them were moved to other dishes on fresh Zvyagintsev's medium, after which were incubated at 30°C. The discs were moved at 20 - 22°C during 15 – 20 minutes. After 30 – 40 min of incubation, the medium beneath the colonies became yellow. In this case, the horizontal profile of the medium color change zones is ring shaped, located directly beneath the colony growth zone. Further incubation increased the ring width, and it transformed into a circle. The results of this experiment prove the suggestion that acids were excreted only or mostly by cells of the growth peripheral zone. When mycelium present in the center of the colony finished developing, its cells stopped excreting acids. However, it is not improbable that moving of the colonies to fresh medium may initiate branching and growth of hyphas accompanied by acid excretion in their centers.

At the constant radial rate, width (ω), cross-section profile and physical density of fungal colonies growth zones formed are also constant [137, 402, 403], and consequently, the mass of the growth zone (M_z) is proportional to its square (G_z), i.e.

$$M_z = k_6 G_z, \tag{A3-9}$$

where k_6 = const is an empirical coefficient. Obviously in this case, the radial growth rate of the central part of the colony is also constant and equal to that of the whole colony and, consequently, the colony central part increase is characterized by equation (A3-2). Contracting the central zone increase growth rate of the colony from that of the whole colony, one may show that the growth zone increase rate is constant:

$$V_{gz} = 2\pi V_r \omega = \text{const}, \tag{A3-10}$$

where V_{gz} is the zone increase rate. In this case, the growth zone mass increase rate is also constant:

$$V_{mz} = k_6 V_{gz} = \text{const}. \tag{A3-11}$$

According to kinetics of any regular process, for example, chemical kinetics of the zero order [280, 392], the growth zone mass will increase linearly with time:

$$M_z = V_{mz} t. \tag{A3-12}$$

Substituting the mass from equation (A3-12) into equation (A3-7) and integrating it, we get:

$$P_2 = qV_{mz}t^2/2 = k_7 t^2, \tag{A3-13}$$

where P_z is the number of acids, which according to the theoretical model should be excreted by the growth zone cells; $k_7 = qV_{mz}/2$ = const.

Equation (A3-13) characterizing expected kinetics of the acid excretion by the growth zone cells is analogous to equation (A3-4) describing their real accumulation in the medium beneath the fungal colonies. As a consequence, excretion of acids by the growth zones of all 9 fungi proceeded with due regard to the theoretical model.

The graphs characterizing *Aspergillus niger* strains in the liquid nutritious media indicate that the quantity of citric acid is increased approximately proportionally to the producer biomass growth [132, 147, 399], i.e. as well as beneath the colonies of 9 above-mentioned fungi. Theoretically, it should be expected that accumulation of the acids should "take the lead" over the biomass growth. This contradiction may be explained by acid excretion not by the whole mycelium, but the hypha end cells only. As not all mycelium cells form branches, the quantity of the end cells is increased relatively slower than the fungus biomass. That is why the acid quantity detected in the medium and the rate of its excretion are lower than theoretically expected ones.

It is assumed that one of the main factors stipulating damaging of various industrial materials by mycelial micromycetes are excreted acids [388]. The data shown indicate that these fungi excrete acids only or mostly during the active growth stage. As shown before, after reduction of more favorable nutritious substrate concentration in the medium, mycelial micromycetes may utilize organic acids, previously excreted by them [393]. All these facts give an opportunity to assume that acids play the role of fungal corrosion factor mostly during the material molding development, and corrosive activity of calm mycelium is lower than that of the growing one.

References

1. GOST 9.102-91. *Unified system preventing corrosion and aging. Impact of biological factors on technical objects. Terms and definitions*. Izd-vo standartov, 1989. (Rus)
2. *Biodamaging. Textbook*. Ed. Prof. V.D. Ilyichev, Moscow, Vyshchaya Shkola, 1987, 352 p. (Rus)
3. Anisimov A.A. and Smirnov V.F., *Biodamages in industry and their elimination*, Gorkyi, GGU Edition, 1980, 82 p. (Rus)
4. Gerasimenko A.A. *Machinery protection from biodamages*, Moscow, Mashinostroenie, 1984, 111 p. (Rus)
5. Osnitskaya L.K., *Miscrobiologia*, 1946, vol. **15**, Iss. 2, pp. 249 – 263. (Rus)
6. Flerov B.K., Biological damages to industrial materials and articles from them, In Coll.: *Problems of Biological Damages and Overgrowths of Materials, Articles and Structures*, 1972, Moscow, Nauka, pp. 1 – 10. (Rus)
7. Skryabina T.G. and Lazareva I.V., Bacterial infectiousness of diesel oils, *Oil processing and petrochemistry (scientific research achievements and progressive experience)*, 1994, No. 6, pp. 14 – 17. (Rus)
8. Lesdbetter E.R. and Foster J.R., *Arch. Microbiol.*, 1960, vol. **35**(2), pp. 104 – 134.
9. Iverson W.P., Microbiological corrosion, *Gas*, 1969, vol. **14**(2), pp. 56 – 60.
10. Poglazova M.N. and Martsevitch I.N., *Miscrobiologia*, 1984, vol. **53**, pp. 850 – 858. (Rus)
11. Foster J.W., *J. Microbiol. A. Serol.*, 1962, vol. **28**(3), pp. 242-287.
12. Rozanova E.P. and Kuznetsov S.I., *Microflora of oil fields*, Moscow, Nauka, 1974, 198 p. (Rus)
13. Krynitsky G.A., *Naval Research Rev.*, 1964, vol. **17**(2), pp. 62-69.
14. Sidorenko L.P. and Dakhnovskaya V.I., Fungal corrosion of metal parts of various devices under marine tropical climate conditions. Biodamages. *Thes. Proc. 2nd All-Union Conference on Biodamages*, Gorky, GGU, 1981, p. 96. (Rus)
15. Hill E.C. *et al.*, *J. Inst. Petrol.*, 1967, vol. **53**, pp. 280 - 284.
16. Hitzmann D.O. and Linnard R.E., *Conf. Petr. VII*, Mexico, 1967, Symp., **36**, pp. 33 – 35.
17. Iverson W.P., Microbial corrosion of iron, In: *Microbial Iron Metabolism. A Comprehensive Treasure*, New York, 1974, pp. 475 – 513.

18. Reviere Jacgues. Microorganisms et carburants accidents et remedes, Petrole et Techn., 1986, No. 322, pp. 73 – 76.
19. Mekhtieva N.A. and Kandinskaya L.I., Spreading of micromycetes in fuel systems of airplanes, *Biological Damages of Construction and Industrial Materials*, Kiev, Naukova Dumka, 1978, pp. 112 – 114. (Rus)
20. Parbery D.G., *Intern. Biodeterior. Bull.*, 1968, vol. **4**(2), pp. 79 – 81.
21. Iverson V.R., *Chem. Eng.*, 1968, vol. **1 - 2**(8), pp. 242 - 244.
22. Iverson V.R., *Corros. Preven. Control*, 1969, vol. **16**(1), pp. 15 – 19.
23. Ermolenko Z.M., Shtuchnaya G.V., and Martovetskaya I.A., Physiological, ultra-structural and morpho-population features of *Mycobacterium flavesceus* bacteria utilizing crude oil and petroleum products, *Biotekhnologia*, 1996, No. 5, pp. 17 – 24. (Rus)
24. Skribachilin V.B., Lapteva E.A., and Mikhailova L.K., On biodamaging of fuels, *Chemistry and Technology of Fuels and Oils*, 1983, No. 12, pp. 29 – 30. (Rus)
25. Litvinenko S.N., *Biological damaging of crude oil and petroleum products and their protection during transportation and storage*, Moscow, Izd. TsNIITEneftekhim, 1970, 51 p. (Rus)
26. Krein S.E., *Applied biochemistry and microbiology*, 1969, vol. **5**, Iss. 2, pp. 233 – 236. (Rus)
27. Krein S.E., *Applied biochemistry and microbiology*, 1973, vol. **9**, Iss. 1, pp. 143 – 145. (Rus)
28. Microflora and crude oil, *Proc. VIII World Oil Congress*, Moscow, Izd. AN SSSR, 1971, 112 p. (Rus)
29. Litvinenko S.N., *Petroleum product protection from impact of microorganisms*, Moscow, Khimia, 1977, 142 p. (Rus)
30. Andreyuk E.I. and Kozlova I.A., *Lithotrophic bacteria and microbiological corrosion*, Kiev, Naukova Dumka, 1977, 162 p. (Rus)
31. Emelin M.I. and Gerasimenko A.A., *Machine protection from corrosion under operation conditions*, Moscow, Mashinostroenie, 1980, 224 p. (Rus)
32. *Radioelectronic equipment protection from external climatic conditions*, Ed. G. Yubish, Izd. Energiya, 1970. (Rus)
33. Zanina V.V., Kopteva A.E., and Kozlova I.A., The effect of biocorrosion soil activity on bioresistance of insulating covers, *Mikrobiologia*, 1996, vol. **58**(1), pp. 88 – 96. (Rus)
34. Romanova N.S. and Zhilina G.S., In: *Novel atmosphere-resistant and decorative covers*, Moscow, F.E. Dzerzhinskii MDNTP, 1976, pp. 126 - 132. (Rus)
35. Proektor E.G., Anastasiev P.I., and Kolyada A.V., *Protection of cables and air electric power supply lines from corrosion*, Moscow, Energiya, 1974, 158 p. (Rus)

36. Mogilnitskii G.M., Zhukova S.V., and Khramikhina V.F., On the problem of bioresistant tests of insulating covers, In: *Determination methods for material bioresistance*, Moscow, VNIIST Minneftegasstroi, 1979, pp. 106 – 112. (Rus)
37. Lugauskas A.Yu., Girigaite L.M., Repechkene Yu.P., and Shlyauzhene D.Yu., Species composition of microscopic fungi and microorganism association on polymeric materials, In: *Actual problems of biodamages*, Moscow, Nauka, 1983. (Rus)
38. Richter M. and Bartakova B., *Tropical adjustment of electrical equipment*, Gosenergoizdat, 1962, 400 p. (Rus)
39. Hedrick N.G., Microbiological corrosion of aluminum, *Mater. Prot.*, 1970, vol. **9**(1), pp. 27 – 31.
40. Parbery D.G., The role of *Cladosporium resinae* in the corrosion of aluminum alloys, *Intern. Biodeter. Bull.*, 1968, vol. **4**(2), pp. 79-81.
41. Zaikina N.A., Elinov N.P., Golovanenko G.G., and Vinogradov P.A., Mould fungi in microbiological damages, *Thes. All-Union Symp.: Theoretical problems of biological damages of materials. AS USSR*, Moscow, 1971, 26 p. (Rus)
42. Microbiologically inflamed corrosion, *Anti-Corros. Meth. and Mater.*, 1994, vol. **41**(6), p. 26.
43. Eating away at the infrastructure – the heavy cost of microbial corrosion, *Water Qual.*, 1995, No. 4, pp. 16 – 19.
44. Zhebrovskii V.V. and Rubinstein F.I., New varnish coatings for countries with tropical climate, In Coll.: *Varnish Coatings for Tropical Climate Conditions*, Profizdat, 1997, No. 1, pp. 19 – 26. (Rus)
45. Blagnik R. and Zanova V., *Microbiological Corrosion*, Moscow, Khimia, 1965, 222 p. (Rus)
46. Zabyirina K.I., Electrically insulating varnishes and materials designed for operations under tropical climate conditions, In Coll.: *Varnish Coatings For Tropical Climate Conditions*, Profizdat, 1977, No. 1, pp. 27 – 44. (Rus)
47. Karyakina M.I. and Maiorova N.V., *Varnishes*, 1985, No. 5, pp. 35 – 37. (Rus)
48. Sukhareva L.A., Semenov G.V., Sergienko T.E., and Yakovlev V.S., Bactericidal corrosion-resistant epoxy coatings of different applications, In Coll.: *Ecological Problems of Biodegradation of Industrial and Construction Materials and Production Waste*, Penza, NII Khimii NGU, 1998, pp. 9 – 12. (Rus)
49. Yamanov S.A., Tropical resistance and tropical protection of electrical insulation materials, *Proc. 1st Inter-High School Conf. on Modern Technology of Dielectrics and Semiconductors*, 1967. (Rus)

50. Sapozhnikova S.A., Some features of tropical and subtropical Asia, In Coll.: *Varnish Coatings For Tropical Climate Conditions*, 1977, No. 1, Profizdat, pp. 4 – 11. (Rus)
51. Lisina-Kulik E.S., Estimation of the impact of some factors on resistance of varnish coatings to fungus damage, *Varnishes and Their Application*, 1971, No. 4, pp. 58 – 61. (Rus)
52. Tsigikalov E. and Sazonov I., At the sea shore, *Tekhnika i Vooruzhenie*, 1968, No. 7, 7 p. (Rus)
53. Lisina-Kulik E.S. and Moiseeva N.G., Survival of fungus spores on varnish coatings for devices in the mode of systematic heating, *Varnishes and Their Application*, 1972, No. 3, pp. 33 – 34. (Rus)
54. Nasonov K.V. and Sharapov V.D., *Ship Conservation*, Sudostroenie, 1972, vol. **17**(31), pp. 133 – 134. (Rus)
55. Blahnick R., Smery soncasnelo vyzkumu v oblasti microbialni koroze naterovych hmot ve svete, *Elektrotech. Klimatotechnol.*, 1963, vol. **4**, pp. 3 - 4.
56. Il'ichev V.D., Bocharov B.V., and Gorlenko M.V., *Ecological Grounds for Protecction Against Biodamages*, Moscow, Nauka, 1985, 62 p. (Rus)
57. Yakubovich D.S., In: *Atmosphere-Resistant Varnishes and Prediction of Their Service Life*, Moscow, F.E. Dzerzhinsky MDNTP, 1982, pp. 7 – 11, 57. (Rus)
58. Nitsberg L., Varnish coatings, *Tekhnika i Vooruzhenie*, 1967, No. 4, pp. 12 – 13. (Rus)
59. *Biological Damage to Construction and Industrial Materials*, Kiev, Naukova Dumka, 1978, 265 p. (Rus)
60. Miller J.D.A. and King R.A., Biodeterioration of metals, In: *Microb. Aspects Deterior. Mater.*, London, 1975, pp. 83 – 103.
61. Pozdneva N.I., Pavlova V.G., Pimenova M.N., and Zinevich A.M., Investigations of polyethylene insulation resistance to impact of microorganisms, *Thes. All-Union Symp. Theor. Problems of Biological Deterioration of Materials*, AN SSSR, Moscow, 1971, pp. 3 - 5. (Rus)
62. Belokon N.F., Tatevosyan E.A., Filatov I.S., and Tsvetkova O.N., Fungus-resistant phenoplasts with organic filler, *Thes. Rep. Second All-Union Symposium on Biological Deterioration and Overgrowth of Materials, Articles and Structures*, Moscow, Nauka, 1972, pp. 54 – 55. (Rus)
63. Rudakova A.K., Microbial corrosion of polymeric materials (poly(vinyl chloride) plasticates and polyethylene) applied in cable industry and methods of its prevention, *Thes. Cand. Diss.*, MSU, 1969. (Rus)
64. Isolation of decomposer fungi with plastic degrading ability, *Philipp J. Sci.*, 1997, vol. 126(2), pp. 117 – 130. (Rus)

65. Belokon N.F., Tatevosyan E.A., and Shidkova G.A., Investigation methods of plastics fungus resistance, *Plasticheskie Massy*, 1974, No. 9, pp. 65 – 67. (Rus)
66. Bilai V.I., Koval E.Z., and Sviridovskaya L.M., Study of fungus corrosion of various materials, In: *Proc. IV Congress on Microbiology of the Ukraine*, Kiev, 1975, p. 85. (Rus)
67. Pankhurst E.S., Protective coatings and wrappings for buried pipes microbiological aspects, *J. Oil and Colour Chem. Assoc. I.*, 1973, vol. **6**(8), pp. 373 – 381.
68. Lugauskas A.Yu. and Stakishaitite R.V., Studying fungi inhabiting materials applied in radio engineering, In: *Biological Deterioration of Materials*, Vilnius, 1979, pp. 72 – 78. (Rus)
69. Huang S.I., The effect of structural variation on the biodegradability of synthetic polymers, *Amer. Chem. Bac. Polym. Prepr.*, 1977, vol. **1**, pp. 438 - 441.
70. Koval E.Z., Likhtenshtein V.I., Likhtenshtein G.A., Glubokaya G.V. *et al.*, Fungus-induced corrosion of glassy materials applied in structures, In: *Biological Damages of Construction and Industrial Materials*, Kiev, 1978, pp. 111 - 112. (Rus)
71. Allakhverdiev G.A., Martirosova T.A., and Gariverdiev R.D., Variation of physicochemical properties of polymers induced by soil organisms, *Plasticheskie Massy*, 1967, No. 2, pp. 17 – 20. (Rus)
72. Zuev Yu.S. and Degteva T.G., *Elastomer Resistance Under Operation Conditions*, Moscow, Khimia, 1986 (1 q.), 18 p. (Rus)
73. Zuev Yu.S., *Elastomer Degradation In Typical Operation Conditions*, Moscow, Khimiz, 1980, 283 p. (Rus)
74. Mudrov O.A., Labutin A.L., Shitov V.S. *et al.*, *New Elastomeric (Polyurethane) Materials In Shipbuilding*, Leningrad, LDNTP, 1979, 20 p. (Rus)
75. Dubok N.N. and Angert L.G., Biodeterioration of rubbers and methods of their protection, In Coll.: *The First All-Union Conference on Biodeterioration*, Moscow, Nauka, 1978, pp. 16—19. (Rus)
76. Rudakova A.K. and Popova T.A., Fungus resistance of cable materials and articles under natural conditions, In Coll.: *Biological Damages Of Construction And Industrial Materials. AS USSR*, Moscow, 1973, No. 4, pp. 72 – 79. (Rus)
77. Ruban G.I., Study of the fungus resistance of some synthetic materials and their protection from mould, *Thes. Rep. Second All-Union Symp. on Biological Deterioration and Overgrowth of Materials, Articles and Structures*, Moscow, Nauka, 1972, pp. 70 – 71. (Rus)

78. Shirokov A.M., *Grounds For Reliability And Operation Of Electronic Equipment*, Minsk, Izd. Nauka i Tekhnika, 1965, pp. 29 - 30. (Rus)
79. Tolmacheva R.N., Tsendrovsky D.V., and Smirnov V.F., Microbiological deterioration of materials and articles applied in radio engineering, In Coll.: *GGU*, 1987, pp. 30 – 34. (Rus)
80. Rodionova M.S., Kariglazova N.B., Ostrovskaya M.A., and Filimonova N.F., Variation of light permeation and scattering coefficients with regard to mould fungus propagation on optical systems, *Optikomekhanicheskaya Promyishlennost*, 1972, No. 2, pp. 62 – 63. (Rus)
81. Ruban G.I. and Slepukhina N.K., Protection of all-climate design articles against fungal infections, Chemical Protection Measures Against Biocorrosion, *Thes. Rep. to Scientific Seminar*, Ufa, 1980, pp. 52 – 57. (Rus)
82. Lebedev E.M., On the problem of microbial destruction of biological optical devices in coastal areas of moderate climate of the USSR, In Coll.: *Problems of Biological Deterioration and Overgrowth of Materials, Articles and Structures*, Moscow, Nauka, 1972, No. 3, pp. 100 – 101. (Rus)
83. Tkhorzhevskii V.P., *Design and Production of Devices for the Countries with Tropical Climate*, Mashinostroenie, 1971, pp. 9 – 85. (Rus)
84. Rodionova M.S., Study of protection methods for silicate glasses, optical parts and lubricants from deterioration by mould fungi, *Cand. Diss. Thes. VIAM*, 1964, 18 p. (Rus)
85. Rodionova M.S. and Razumovskaya Z.G., On spreading of mould fungi on the surface of optical glasses, In Coll.: *Problems of Biological Deterioration and Overgrowth of Materials, Articles and Structures*, Moscow, Nauka, 1972, No. 3, pp. 79 – 91. (Rus)
86. Jaton C., Attague des pierres calcaires et des betons, *Degradatation Microbienne Material*, Paris, 1974, pp. 41 – 56.
87. Kaller A., *Fungysbilding auf Optik*, Feingeratetech nik., 1960, No. 1.
88. *Reliability of Surface Radio and Electronic Equipment*, Sovetskoe Radio, 1987, pp. 54 – 306. (Rus)
89. Extrapolation of biodegradability test data by use of growth kinetic parameters, *Ecotoxical and Environ. Safety*, 1994, vol. **27**(3), pp. 306 – 315.
90. Petinov O.V. and Shcherbakov E.F., Tests of electrical devices, *The Manual for the Higher School*, Moscow, Vyisshaya Shkola, 1985, 214 p. (Rus)
91. Wasserbauer R., Czechoslovak research into microbiological corrosion of electrical equipment, *Intern. Biodeter. Bull.*, 1967, vol. **3**(1), pp. 1 – 2.
92. Ruban G.I., Microscopic fungi damaging articles of electronic equipment, *Elektronnaya Tekhnika*, 1976, vol. 8(8-50), pp. 82 - 88. (Rus)
93. Kuznetsov S.I., The value of sulfate-restoring bacteria for metal equipment corrosion, In: *Theory and Practice of Anti-corrosion Protection of Underground Structures*, Moscow, AN SSSR, 1958, pp. 246 - 251. (Rus)

94. Tiller A.R. and Booth G.H., Anaerobic corrosion of aluminum by sulfate-reducing bacteria, *Corros. Sci.*, 1968, vol. **8**, pp. 549 – 555.
95. Gattelier C.R., Pollution microbienne des carburants et corrosion des reservoirs, *Corrosion* (France), 1973, vol. **21**(2), pp. 103 – 109.
96. Leonard I.M., *Failure of Electronic Equipment under Tropical Service Division U.S.*, Naval Research Laboratory, Washington, D.C., February, 1985.
97. Belousov L.K. and Savchenko V.S., Electrical connectors in radio and electronic equipment, *Energia*, 1967, pp. 9 - 12. (Rus)
98. Tvorzhevskii V.P. and Perevezentsev I.G., *Designing of Devices for the Countries with Tropical Climate*, Moscow, Mashgiz, 1960, 154 p. (Rus)
99. Shirokov A.M., *Reliability of Radio Electronic Devices*, Izd. Vysshaya Shkola, 1972, pp. 62 – 64. (Rus)
100. Mishustin E.N., Pertsovskaya M.I., and Gorbov V.A., *Sanitary Microbiology of Soils*, Moscow, Nauka, 1979, 265 p. (Rus)
101. Babieva I.P. and Agre N.S., *Practical Guide on Soil Biology*, Moscow, Izd. MGU, 1971, 140 p. (Rus)
102. Overgrowth and biodeterioration. Environmental problems, In: *Scientific Proceeding Collection*, Moscow, Nauka, 1992, vol. **12**(5), 252 p. (Rus)
103. Aristovskaya T.V., *Microbiology of Soil Formation Processes*, Moscow, Nauka, 1980, 187 p. (Rus)
104. Nikitina Z.I., *Microbiological Monitoring of Surface Ecosystems*, Novosibirsk, Nauka, 1991, 221 p. (Rus)
105. Kozhevin P.A., *Microbial Populations in Nature*, Moscow, Izd. Universiteta, 1989, 173 p. (Rus)
106. *Experimental Microbiological Methods, The Reference Book*, Ed. V.I. Bilai, Kiev, Naukova Dumka, 1982, 550 p. (Rus)
107. Methods of Soil Microbiology and Biochemistry, *The Manual for Students of Universities*, Moscow, MSU, 1991, 303 p. (Rus)
108. Kuznetsov S.I. and Dubinina G.A., *Investigation Methods for Aquatic Microorganisms*, Moscow, Nauka, 1989. (Rus)
109. Zvyagintsev D.G., *Interaction of Microorganisms with Solid Surfaces*, Moscow, Izd. MGU, 1973, 47 p. (Rus)
110. Zvyagintsev D.G., *Microbiological Sciences*, 1967, No. 3, 97 p. (Rus)
111. Deryagin B.V. and Krotova N.A., *Adhesion*, Moscow, Izd. AN SSSR, 1979, 246 p. (Rus)
112. Zimon A.D., *Adhesion of Dust and Powder*, Moscow, Izd, AN SSSR, 1979, 246 p. (Rus)
113. Zvyagintseva I.S. and Zvyagintsev D.G., The effect of microorganism cell adsorption on transformation of some steroids, *Biologicheskaya Nauka*, 1998, No. 5, pp. 20 – 28. (Rus)

114. Arkhipenko V.I., Gerbilskii L.V., and Cherchenko Yu.P., Structure and functions of intercellular constants, In: *Structure and Functions of Biological Membranes*, Moscow, Nauka, 1975, 347 p. (Rus)
115. Bobkova T.S., Chekunova L.N., and Zlachevskaya I.V., Fungus adhesion on vitroceramics, *Mikrobiologia i Fitopatologia*, 1979, vol. **13**(3), pp. 208 – 213. (Rus)
116. Dijkerman Rembrandt, Vervuren Mike B, Op Den Camp Huub J.J., and van der Drift Chris, Adsorption characteristics of cellulolytic enzymes from the anaerobic fungus *Piromyces* sp. strain E2 on microcrystalline cellulose, *Appl. and Environ. Microbiol.*, 1996, vol. **62**(1), pp. 20 – 25.
117. Starostina N.G., Koshchaev A.G., and Ratner E.N., Hydrophoby of the cell surface of methanotrophic bacteria: comparative estimation and applied aspects. Autotrophic microorganisms, *Proc. Conf. on Remembrance of RAS Acad. Kondratieva E.N.*, Moscow, MSU, Biological Department, April 23 - 25, 1996, p. 92. (Rus)
118. Kitano T., Yutani Y., Shimazu A., Yano I., Ohashi H., and Yamano J., The role of physicochemical properties of biomaterials and bacterial cell adhesion *in vitro*, *Int. Artif. Organs*, 1996, vol. **19**(6), pp. 353 - 358.
119. Pisarev O.A., Samsonov G.V., Bernotaitite M.V., and Muravieva T.D., Full reversibility of sorption of biologically active substances by strongly cross-linked gel carboxyl cation exchangers, *Prikladnaya Biokhimia i Microbiologia*, 1996, vol. **32**(3), pp. 280 – 283. (Rus)
120. Kisten A.G., Kigel N.P., Kurdish I.K., and Gordienko A.S., Influence of some physicochemical environmental factors on adhesion of metatrophic bacteria, *Mikrobiologichesky Zhurnal*, 1996, vol. **58**(3), pp. 62 – 70. (Rus)
121. Lindqvist Roland and Bengtsson Gorau, Diffusion-limited and chemical-interaction-dependent sorption of soil bacteria and microspheres, *Soil Biol. and Biochem.*, 1995, vol. **27**(7), pp. 941 – 948.
122. Penalver Carmen M., Casanova Manuel, Martinez Jose P., Cell wall protein and glycoprotein constituents of *Aspergillus fumigatus* that bind to polystyrene may be responsible for the cell surface hydrophobicity of the mycelium, *Microbiology*, 1996, vol. **142**(7), pp. 1597 – 1604.
123. Jansen B. and Kohnen W., Prevention of biofilm formation by polymer modification, *J. Ind. Microbiol.*, 1995, vol. **15**(4), pp. 391 – 396.
124. James G.A., Beandette L., and Costerton J.W., Interspecies bacterial interactions in biofilm, *J. Ind. Microbiol.*, 1995, vol. **15**(4), pp. 257 – 262.
125. Patti Joseph M. and Allen Bradley L., Mscramm—ediated adherence of microorganisms to host tissues, *Ann. Rev. Microbiol.*, 1994, vol. **48** – Palo Alto (Calif.), pp. 585 – 617.

126. Kornev N.R., Karaev Z.O., and Soldatenko N.K., Electrokinetic characteristic of *Candida albianc* fungus cell surface, *Mikologia i Fitopatologia*, 1985, vol. **19**(6), pp. 490 – 494. (Rus)
127. Fletcher M. and Floodgate I.D., An electron microscope demonstration of an acidic polysaccharide in the adhesion of a marine bacterium to solid surfaces, *J. Gen. Microbiol.*, 1973, vol. **74**, pp. 325 – 334.
128. Vtyutin B.V. and Paltsyin A.A., *The Modern Methods and Techniques of Electron Microscopic Investigations of Biological Objects*, Moscow, Radio i Svyaz, 1985, 56 p. (Rus)
129. Michell A.J. and Scurfield J., An assessment of infrared spectra as indicators of fungal cell wall composition, *Austral. J. Biol. Sci.*, 1970, vol. **23**(2), pp. 345 – 360.
130. *Microbiological Adhesion and Aggregation*, Ed. K.C. Marchall, Berlin etc., Springer-Verlag, 1984, p. 124.
131. Gulyaeva N.D., Zaslavskii B.Yu., Rogozhin S.V., and Lyapunova T.S., Application of infrared microscopy to studies of yeast cell wall chemical composition, *Mikrobiologia*, 1977, vol. **46**(4), pp. 667 – 671. (Rus)
132. Shelegel G., *Allgemeine Microbiologie*, Georg Thieme Verlag, Stuttgard, 1985.
133. *Smaller Practicum on the Lowest Plants*, Moscow, Vysshaya Shkola, 1976, 216 p. (Rus)
134. Bilai V.I., *Grounds for General Mycology*, Kiev, Vysshaya Shkola, 1980, 360 p. (Rus)
135. Braun V., *Bacterial Genetics*, Ed. Alikhanyan, Moscow, Nauka, 1968, 446 p. (Rus)
136. Bilai V.I., *Biologically Active Substances of Microscopic Fungi and Their Application*, Kiev, Naukova Dumka, 1965, 267 p. (Rus)
137. Pirt S.J., A kinetic study of the mode of growth of surface colonies of bacteria and fungi, *J. Gen. Microbiol.*, 1967, vol. **47**(1), pp. 181 – 197.
138. Baskanyan I.A., Birsonov V.V., and Kryilov Yu.M., Mathematical description of main kinetic regularities of microorganism cultivation, Mikrobiologia, 1976, vol. **5**, Iss. 1, pp. 5 - 75. (Rus)
139. Tsiperovich A.S., *Enzymes*, Kiev, Tekhnika, 1971, 354 p. (Rus)
140. Bilai T.I., Enzymatic processes in biocorrosion, In: *Biological Deterioration of Construction and Industrial Materials*, Kiev, Naukova Dumka, 1978, pp. 68 - 69. (Rus)
141. Orlova E.N., Fungal utilization of polymeric materials, *Mikologia i Fitopatologia*, 1980, vol. **14**, iss. 5, pp. 422 – 425. (Rus)
142. Tirpak G., Microbial degradation of plasticized P.V.C., *Sp. Journal*, 1970, vol. **26**, p. 26.

143. Bilai V.I., Lizak Yu.V., Novikova T.N., and Koval E.Z., Cellulosolytic properties of fungi damaging graphics, *Mikrobiol. Zh. AN UkrSSR*, 1978, vol. **40**(5), p. 577. (Rus)
144. Zagulyaeva Z.A., Some data on cellulose degradation by micromycetes, In: *Problems of Biological Deterioration and Overgrowth of Materials, Articles and Structures*, Moscow, Nauka, 1972, pp. 51 – 54. (Rus)
145. Nyuksha Yu.P., Rapid determination of cellulose-based material fungal resistance, In: *Biological Deterioration of Construction and Industrial Materials*, Kiev, Naukova Dumka, 1978, pp. 158 – 164. (Rus)
146. Turkova Z.A., Deterioration of some technical materials by fungi, In: *Biocorrosion, Biodeterioration, Overgrowth*, Moscow, 1976, pp. 71 – 80. (Rus)
147. Fenixova R.V., Enzyme biosynthesis by microorganisms, In: *Problems of Biological Deterioration and Overgrowth of Materials, Articles and Structures*, Moscow, Nauka, 1973, pp. 5 – 10. (Rus)
148. Eriksson K.-C. and Larsson K., Fermentation of waste mechanical fibers from a newsprint mill by the rot fungus *Sporotrichum pulverulentum*, *Biotechnol. and Bioeng.*, 1975, vol. **17**(3), pp. 137 – 348.
149. Rosenberg S.L., Cellulose and lignocellulose degradation by thermophilic and thermotolerant fungi, *Mycologia*, 1978, vol. **LXX**(1), pp. 1 – 13.
150. Turkova Z.A., Mycoflora of mineral-based materials and probable mechanisms of their degradation, *Micologia i Fitopatologia*, 1974, vol. **8**(3), pp. 219 – 226. (Rus)
151. Vasnev V.A., Biodegradable polymers, *Vysokomol. Soedin., Ser. B*, 1997, vol. **39**(13), pp. 2073 – 2086. (Rus)
152. Huang S.J., The effect of structural variation on the biodegradability of synthetic polymers, *Amer. Chem. Bacteriol. Polym. Prepar.*, 1977, vol. **1**, pp. 438 – 441.
153. Borrow A., The metabolism of *Gibberrella fugikuroi* in stirred culture, *Can. J. Microbiol.*, 1961, vol. **7**(2), p. 227.
154. Bu Lock I.D., Regulation of 6-methyl-salicylate and patulin synthesis in *Penicillium urticae, Can J. Microbiol.*, 1964, vol. **15**(3), p. 279.
155. Caldwell I.V. and Trinci A.P., The growth unit of the mould *Georichum candidum, Arch. Microbiol.*, 1973, vol. **88**(1), pp. 1 – 10.
156. Varfolomeev S.D. and Kalyazhny S.V., Biotechnology: Kinetic Grounds of Microbiological Processes, *Handbook for Biology and Chemistry Specialized Higher Schools*, Moscow, Vysshaya Shkola, 1990, 296 p. (Rus)
157. Grove S.N., Bracker C.E., and Marre D.J., An ultrastructure basis for hyphal tip growth in *Pythium ultimum, Amer. J. Bot.*, 1975, vol. **59**(2), pp. 245 – 266.

158. Gromov B.V. and Pavlenko G.V., *Bacterial Ecology*, Handbook, Leningrad, Izd. Leningradskogo Universiteta, 1989, 248 p. (Rus)
159. Ierusalimskii N.D., *Microbe Physiology Bases*, Moscow, Izd. AN SSSR, 1963, 244 p. (Rus)
160. Ierusalimskii N.D. and Neronova N.N., Quantitative dependence of microorganism growth rate on concentration of metabolites, *Doklady AN SSSR, Ser. Biol.*, 1965, vol. **161**(6), pp. 1437 – 1440. (Rus)
161. Baskanyan I.A., Biryukov V.V., and Kryilov Yu.M., Mathematical description of basic kinetic regularities of microorganism cultivation, *Mikrobiologia*, 1976, vol. **1**(5), pp. 5 – 75. (Rus)
162. *Technical Mycology*, Ed. I. Ya. Veselov, Moscow, 1972. (Rus)
163. Emanuel N.M., *Kinetics of Experimental Tumor Processes*, Moscow, Nauka, 1977, 356 p. (Rus)
164. Kulik E.S., Karyakina M.I., Vinogradova L.M., and Moiseeva N.G., The role of fungal ecology investigations in determination of fungal resistance of varnishes and paints, In: *Microorganisms and Lower Plants – the Degraders of Materials and Articles*, Moscow, Nauka, 1979, pp. 90 – 96. (Rus)
165. Ruban E.L., *Microbial Lipids and Lipases*, Moscow, Nauka, 1977, 215 p. (Rus)
166. Abramova N.F., Naplekova N.N., and Shkulova G.A., Physiological activity of mould fungus cultures during growth on plastics and its variation with regard to storage method, In: *Biological Deterioration of Construction and Industrial Materials*, Kiev, Naukova Dumka, 1978, pp. 69 – 70. (Rus)
167. Naplekova N.N. and Abramova N.F., Microbiological deterioration of plastics, *Izv. SO AN SSSR, Ser. Biol.*, 1978, No. **15**(3), pp. 42 – 47. (Rus)
168. Huang C. and Yannas I.V., Mechanochemical studies of enzymatic degradation of insoluble collagen fibers, *J. Biomed. Mater. Res.*, 1984, No. 8, pp. 137 – 154.
169. Gumargalieva K.Z., Moiseev Yu.V., and Zaikov G.E., Macrokinetic aspects of biocompatibility and biodegradability of polymers, *Uspekhi Khimii*, 1994, vol. **63**(10), pp. 905 – 921. (Rus)
170. Mlinac M. and Munjko I., Degradactja biorargradljvoy polietilena niske gustoce, *The 2nd Intern. Symp. Degradation and Stabilization of Polymers*, 1978, pp. 100 - 101. (Rus)
171. Zuev Yu.S., *Polymer Degradation Caused by Aggressive Media*, Moscow, Khimia, 1972, 229 p. (Rus)
172. Tyinnyi A.N., *Strength and Destruction of Polymers Caused by Liquids*, Kiev, Naukova Dumka, 1975, 247 p. (Rus)

173. Manin V.N. and Gromov A.N., *Physicochemical Resistance of Polymeric Materials Under Operation Conditions*, Leningrad, Khimia, 1980, 248 p. (Rus)
174. Vorobieva G.Ya., *Chemical Resistance of Polymeric Materials*, Moscow, Khimia, 1981, 296 p. (Rus)
175. Moiseev Yu.V. and Zaikov G.E., *Chemical Resistance of Polymers in Aggressive Media*, Moscow, Khimia, 1979, 287 p. (Rus)
176. Emanuel N.M. and Buchachenko A.L., *Chemical Physics of Aging and Stabilization of Polymers*, Moscow, Nauka, 1982, 359 p. (Rus)
177. Zhuk N.P., *The Course of Metal Corrosion and Protection Theory*, Moscow, Metallurgia, 1976, 472 p. (Rus)
178. Deshelev S.F. and Gurvich F.G., *Expert Assessment of Mathematical Statistic Methods*, Moscow, Statistika, 1980, 262 p. (Rus)
179. Mirkin B.G., *Group Choice Problems*, Moscow, Nauka, 1974, 253 p. (Rus)
180. *Corrosion, Aging and Biodeterioration Protection of Machines, Equipment and Structures*, Reference book (two volumes), Ed. A.A. Gerasimenko, Moscow, Mashinostroenie, 1987, 688 p. (Rus)
181. Semenov S.A., Ozhegin F.F., Ryizhkov A.A., and Ivanov S.S., The effect of organic acids on mechanical properties of poly(methyl methacrylate), *Fiz.-Khim. Mekhanika Materialov*, 1986, No. 5, pp. 120 – 122. (Rus)
182. Semenov S.A., Gerasimenko A.A., Zhdanova O.A., and Ryizhkov A.A., Forecasting of microbiological degradability of BPVL type wires, In: *Weapon Protection from Corrosion, Aging and Biodeterioration. KNS MO Materials*, Vedomstvennoe Izdatelstvo, 1988, pp. 165 – 178. (Rus)
183. *GOST 9.048-89. Uniform System of Corrosion and Aging Protection. Technical Articles. Laboratory Test Methods for Resistance to Mould Fungus Impact.* Moscow, Izd. Standartov, 1989. (Rus)
184. *GOST 9.049-91. Uniform System of Corrosion and Aging Protection. Polymeric Materials. Laboratory Test Methods for Resistance to Mould Fungus Impact*, Moscow, Izd. Standartov, 1991. (Rus)
185. *GOST 9.082-77. Uniform System of Corrosion and Aging Protection. Oils and Lubricants. Laboratory Test Methods for Resistance to Bacterium Impact*, Moscow, Izd. Standartov, 1977. (Rus)
186. *GOST 9.023-74. Uniform System of Corrosion and Aging Protection. Laboratory Test Methods for Biological Resistance of Fuels Protected by Anti-bacterial Additives*, Moscow, Izd. Standartov, 1974. (Rus)
187. Semenov S.A., Gorshkov V.A., and Gumargalieva K.Z., *Quantitative Estimation of Biosoiling of Polymeric Materials and Components of Weapons by Technophilic Microorganisms. Methodological Guidelines*, vol. **6221**, Vedomstvennoe Izdatelstvo, 1990, 76 p. (Rus)

188. Zhdanova O.A., Semenov S.A., and Skribachilin V.B., *Detection and Prevention of Biodamages to Insulation Materials of Weapon Electric Circuits. Methodological Guidelines*, vol. **6216**, Vedomstvennoe Izdatelstvo, 1991, 48 p. (Rus)
189. Vasilenko V.T. and Chernenko N.S., *Influence of Operation Factors on Fueling System of Aircraft*, Moscow, Mashinostroenie, 1986, 184 p. (Rus)
190. Skribachilin V.B., On the reasons of soiling of fuel system filters of aircrafts, *Flight Safety Problems*, 1980, No. 4, pp. 41 – 44. (Rus)
191. Skribachilin V.B., Lapteva E.A., Mikhailova L.K., and Semenov S.A., *A Complex of Methods for Estimation of Biological Pollution of Fuels and Lubricants*, vol. **5314**, Vedomstvennoe Izdatelstvo, 1985, 44 p. (Rus)
192. Skribachilin V.B., Lapteva E.A., Mikhailova L.K., and Semenov S.A., *Detection, Elimination and Prevention of Biological Pollution of Fuels and Lubricants during Storage and Operation of Weapons. Methodological Guidelines*, vol. **5476**, Vedomstvennoe Izdatelstvo, 1986, 20 p. (Rus)
193. Naumov N.A. and Kozlov V.E., *Grounds for Botanic Microtechnology*, Moscow, Sov. Nauka, 1954, 312 p. (Rus)
194. Mazola B.A., Melnikov V.A., and Shilov A.G., *Development and Manufacture of Science Intensive Equipment for Genetic Engineering, Biotechnological and Medical and Biological Investigations, Scientific and Applied Developments: Genetics, Selection, Biotechnology*, Novosibirsk, Inst. Cytol. Genet. SO RAS, 1997, 86 p. (Rus)
195. Roskin G.I., *Microscopic Technique*, Moscow, Sovetskaya Nauka, 1967, 447 p. (Rus)
196. Grechushkina N.N. and Nette I.T., Growth of microorganisms in media with petroleum oils, *Vestnik Moskovskogo Universiteta, Ser. Biol. i Pochv.*, 1968, No. 2, pp. 122 – 124. (Rus)
197. Karnaukhov V.N., *Luminescent Spectral Analysis of a Cell*, Moscow, Nauka, 1978, 209 p. (Rus)
198. Lanetsky V.P., Filin-Koldakov B.V., and Pospelova L.P., Detection of a fungal infection in herbal tissues by luminescent microscopy method, *Mikologia i Fitopatologia*, 1976, vol. **10**(6), pp. 516 - 518. (Rus)
199. Meisel M.N. and Glutkina A.V., Luminescent microscopy application for rapid detection of pathological changes in tissues and organs, *Doklady AN SSSR*, 1953, vol. **3**(91), pp. 647 - 657. (Rus)
200. Morshchakova G.N., Biological degradation of oil and petroleum products polluting soils and water, *Biotekhnologia*, 1998, No. 1, pp. 85 - 92. (Rus)
201. Belyakova L.A., On microorganisms developing on aviation materials and fuels, *Biological Damaging of Materials*, Vilnius, 1979, pp. 28 - 32. (Rus)

202. Mekhtieva N.A. and Kandinskaya L.I., Micromycete spreading in fuel systems of aircrafts, *Biological Damaging of Construction and Industrial Materials: Proc. All-Union School-Seminar*, Kiev, 1978, pp. 112 - 114. (Rus)
203. Thomas A.H. and Hill E.S., *Aspergillus fumigatus* and supersonic aviation. 2. Corrosion, *Int. Biodeterior. Bull.*, 1976, vol. **12**(4), pp. 116 – 119.
204. Gibbs C.P. and Davies S.I., The rate of microbial degradation of oil in a beach gravel column, *Microbiol. Ecol.*, 1976, vol. **3**(1), pp. 55 - 64.
205. Kalaganov V.A., Kagan B.I., and Zakirova V.Z., Determination of microbial damage degree for lubricating and cooling fluids, *Khimia i Tekhnologia Topliv i Masel (Chemistry and Technology of Fuel Oils and Oils)*, 1985, No. 8, pp. 44 - 46. (Rus)
206. Parbery D.J., Biological problems in jet aviation fuel and the biology of Amorphothecs resinae, *Mater. Org.*, 1971, vol. **6**, pp. 161 – 208.
207. Vishnyakova T.P., Rabotnova I.L., Grechushkina N.N. et al., Distillate oil fuel protection from microorganisms, In: *Biocorrosion, Biodamage, Overgrowth*, Moscow, 1976, pp. 83 - 86. (Rus)
208. Rabotnova I.L., Vishnyakova T.P., Grechushkina N.N. et al., A technique of laboratory tests of anti-microbial activity of additives to petroleum fuels, In: *Biological Damages of Construction and Industrial Materials*, Moscow, 1973, pp. 58 - 68. (Rus)
209. Egorov N.S., Vishnyakova T.P., Grechushkina N.N. et al., The damage level of distillate petroleum fuels by microorganism and their protection, In: *Biological Damages of Construction and Industrial Materials: Proc. All-Union School-Seminar*, Kiev, 1978, pp. 136 - 137. (Rus)
210. Arutyunov V.D., Production of stable preparations for luminescent microscopy, *Zhurn. Obshch. Biologii*, 1956, vol. **17**(1), pp. 79 - 83. (Rus)
211. Edmons P., Selection of test organisms for use in evaluating microbial inhibitoro in fuel-water systems, *Appl. Microbiol.*, 1965, vol. **13**(5), pp. 823 – 824.
212. Miller T.Z. and Jonson M.X., Utilization of gas oil by yeast culture, *Biotechnol. Bioans.*, 1966, vol. **8**(4), pp. 567 – 580.
213. Epifanova O.I. and Terpsikh, *Radiography Method in the Study of Cellular Cycles*, Moscow, Nauka, 1969, 119 p. (Rus)
214. Ilkov A.T., Isotope methods in microbiology, In: *Experimental Microbiology*, Ed. S.V., Byirdarov, Sofia, Meditsina i Fizkultura, 1965, pp. 166 - 195. (Rus)
215. Hill E.C., Biodegradation of petroleum products, In: *Microbial Aspects of the Deterioration of Materials*, London, 1975, pp. 127 – 136.
216. Ruban E.L., Verbina N.M., and Buzhenko S.A., *Biosynthesis of Amino Acids by Microorganisms*, Moscow, Nauka, 1968, 293 p. (Rus)

217. Asatiani V.S., *Methods of Biochemical Investigations*, Moscow, Medgiz, 1956, 472 p. (Rus)
218. Lowry O.H., Rosenbrough H.J., and Furr A.L., Protein measurement with the Folin phenol reagent, *J. Biol. Chem.*, 1951, vol. **193**(1), pp. 265 – 275.
219. *Patent PCT C 12Q1/04 WO 9504157 A1*. The method of microorganism identification with the help of at least two chromogens (PCT), February 09, 1995. (Rus)
220. Kondakova N.V., Bozhko T.V., Korzhova L.P. et al., Estimation of various quantitative determination methods of native and formaldehyde-processed proteins, *Voprosy Med. Khimii*, 1983, vol. **29**(2), pp. 134 – 140. (Rus)
221. Fazokas S., Webeter R.G., and Datymer A., Two new staining procedures for quantitativeestimation of proteins on electrophoretic stripes, *Biochim. Biophys. Acta*, 1963, vol. **71**(2), pp. 377 – 391.
222. Kazitsyina L.A. and Kupetskaya N.B., *Application of UV-, IR-, NMR-, and Mass-Spectroscopy to Organic Chemistry*, Moscow, Izd. Moskovskogo Universiteta, 1979, 240 p. (Rus)
223. Chernogryadskaya N.A., Rozanov Yu.M., Bogdanova M.S., and Borovikov Yu.S., *Ultraviolet Fluorescence of the Cell*, Leningrad, Nauka, 1978, 215 p. (Rus)
224. Ripphahn J. and Halpasp H., Quantitation in high-performance microthin layer chromatography, *J. Chromatogr.*, 1975, vol. **112**, pp. 81 - 96. (Rus)
225. Akhrem A.A. and Kuznetsova A.I., *Microthin Layer Chromatography*, Moscow, Nauka, 1964, 175 p. (Rus)
226. Determann H., Wleland T., and Lubon G., Neue anwendungen der Dunnochlchtchromatographte, *Experimentia*, 1962, Bd. **18**(9), S. 430 – 432.
227. Lurier A.A., *Chromatographic Materials*, Moscow, Khimia, 1978, 440 p. (Rus)
228. Pomortseva N.V., Nette I.T., and Liber L.I., B_6 vitamin formation by Cladosporium resinae, *Appl. Biochemistry and Microbiology*, 1977, vol. **13**(5), pp. 718 - 721. (Rus)
229. Belozersky A.N. and Proskuryakov N.I., *Practical Guides to Biochemistry of Plants*, Moscow, 1951, 388 p. (Rus)
230. *The Method of Biochemical Analysis of Plants*, Leningrad, Izd. Leningr. Univ., 1978, 192 p. (Rus)
231. Yamada K. and Torigoe J., Utilization of hydrocarbons by microorganisms: screening of alkane assimilating fungi and their oxidation products from alcans, *J. Agr. Chem. Soc.*, 1966, vol. **40**(3), pp. 364 – 370.

232. Lin H.E., Jida N., and Ilsuca H., Formation of organic acids and ergosterol from n-alkanes by fungi isolated from oil fields in Japan, *J. Ferment. Technol.*, 1971, vol. **49**(3), pp. 771 – 777.
233. *A Technique for High-quality Paper Chromatography of Saccharums, Organic and Amino Acids*, Ed. O.A. Semikhatov, Leningrad, Izd. AN SSSR, 1962, 85 p. (Rus)
234. Pleshkov B.N., *Practicum on Biochemistry of Plants*, Moscow, Kolos, 1976, 256 p. (Rus)
235. Brovko L.Yu., Ugarova N.N., Vasilieva T.E., Dombrovsky V.A., and Berezin I.V., Bioluminescence, *Biochemistry*, 1978, No. 5, 43 p. (Rus)
236. *Practical Training on Microbiology*, Ed. Prof. N.S. Egorov, Moscow, MSU, 1986, 356 p. (Rus)
237. Kretovich V.L., *Grounds of Biochemistry of Plants*, Moscow, Vysshaya Shkola, 1964, 380 p. (Rus)
238. Pimenova M.N., Grechushnikova N.N., and Azova L.G., *Guidelines for Practical Training on Microbiology*, Moscow, Izd. Mosk. Univ., 1971, 221 p. (Rus)
239. Tsiperovich A.S., *Enzymes*, Kiev, Tekhnika, 1971, 354 p. (Rus)
240. Levina L.Sh. and Lebedeva E.I., Determination of glucose oxidase activity by an express-method, *Nauch. Tekhcn. Inform. – Vinodelch. Promyishlennost*, 1966, vol. **4**, pp. 6 - 9. (Rus)
241. Hill E.C., A simple rapid microbiological test for aircraft fuel, *Aircraft Eng.*, 1970, No. 7, pp. 24 – 26.
242. Kursanov L.I., *A Manual for Determination of Aspergillus and Penicillium Genus Fungi*, Moscow, Medgiz, 1947, 174 p. (Rus)
243. Andreyuk E.I., Bilai V.I., Koval' E.Z., and Kozlova I.A., *Microbiological Corrosion and Its Exciters*, Kiev, Naukova Dumka, 1980, 288 p. (Rus)
244. *Guidelines for Practical Training on Microbiology*, Ed. N.S. Egorov, Moscow, Izd. Moskovskogo Universiteta, 1971, 451 p. (Rus)
245. Anisimov A.A., Semicheva A.S., Aleksandrova I.F., Feldman M.S., and Smirnov V.F., Biochemical aspects of industrial material protection from damaging by microorganisms, In: *Actual Problems of Biodamages*, Moscow, 1983, pp. 77 - 101. (Rus)
246. *GOST 9.053-75. Uniform System of Corrosion and Aging Protection. Non-metal Materials and Articles with Their Application. A Test Method for Microbiological Resistance under Natural Conditions in the Atmosphere.* Moscow, Izd. Standartov, 1975. (Rus)
247. *GOST 16350-80. Climate of the USSR. Zoning and Statistic Parameters of Climatic Factors for Technical Purposes.* Moscow, Izd. Standartov, 1980. (Rus)

248. Braga V.G. and Goroshchenko B.T., *Dynamics of Aircraft Flight*, VVIA, 1968, pp. 48 - 55. (Rus)
249. *The Reference Book of Aviation Technician*, Voenizdat, 1961, 176 p. (Rus)
250. *Tables of Gas Dynamic Values*, TsAGI, 1965. (Rus)
251. Jones G.W., *Adhesion to Animal Surfaces, Microbial Adhesion and Aggregation*, Ed. K.C. Marshall, Berlin, 1984, pp. 303 – 321.
252. Calleja G.B., *Aggregation. Group report, Microbial Adhesion and Aggregation*, Ed. K.C. Marshall, Berlin, 1984, pp. 303 – 321.
253. Rutter P.R., *Mechanisms of Adhesion. Group Report, Microbial Adhesion and Aggregation*, Ed. K.C. Marshall, Berlin, 1984, pp. 5 – 19.
254. Kjelleberg S., *Adhesion to Inanimate Surfaces, Microbial Adhesion and Aggregation*, Ed. K.C. Marshall, Berlin, 1984, pp. 51 - 70.
255. Caldwell D.E., Surface colonization parameters from cell density and distribution, In: *Microbial Adhesion and Aggregation*, Ed. K.C. Marshall, Berlin, 1984, pp. 125 – 136.
256. Fletcher M., Comparative physiology of attached and free-living bacteria, In: *Microbial Adhesion and Aggregation*, Ed. K.C. Marshall, Berlin, 1984, pp. 223 – 232.
257. Atkinson B., Consequences of aggregation, In: *Microbial Adhesion and Aggregation*, Ed. K.C. Marshall, Berlin, 1984, pp. 351 – 371.
258. Amed R. and Bayer E.A., *Experimentia*, 1986, vol. **42**(1), pp. 72 – 73.
259. Breznak J.A., Group report. Activity on surfaces, In: *Microbial Adhesion and Aggregation*, Ed. K.C. Marshall, Berlin, 1984, pp. 203 – 221.
260. Kornev N.R., Kraev N.R., and Soldatenko N.K., Electrokinetic characterization of fungus cell surface, *Mikrobiologia i Fitopathologia*, 1985, vol. **19**(6), pp. 480 – 494. (Rus)
261. Characklis W.G., Biofilm development: A process analysis, In: *Microbial Adhesion and Aggregation*, Ed. K.C. Marshall, Berlin, 1984, pp. 137 – 157.
262. White D.C., Chemical characterization of films, In: *Microbial Adhesion and Aggregation*, Ed. K.C. Marshall, Berlin, 1984, pp. 159 - 176.
263. Rutter P.R. and Vincent B., Physicochemical interactions of the substratum microorganisms and the fluid phase, In: *Microbial Adhesion and Aggregation*, Ed K.C. Marshall, Berlin, 1984, pp. 21 – 38.
264. Robb I.D., Stereo-biochemistry and function of polymers, In: *Microbial Adhesion and Aggregation*, Ed K.C. Marshall, Berlin, 1984, pp. 39 - 49.
265. Marshall K.S., Mechanism of the initial events in the sorption of marine bacteria to surfaces, *General Microbiol.*, 1971, vol. **68**, pp. 337 – 348.
266. Fletcher M., Influence of substratum characteristics on the attachment of a marine Pseudomonad to solid surfaces, *Appl. Environ. Microbiol., Jap.*, 1979, pp. 67 – 72.

267. Gerasimov Ya.I., *The Course of Physical Chemistry*, Moscow, Khimia, 1973, 612 p. (Rus)
268. Ruban G.I. and Slepukhina N.K., Protection of articles of non-climatic design from fungal damages, In: *Chemical Protection Facilities from Biocorrosion*, Ufa, VNIIKhSZR, 1980, pp. 52 – 54. (Rus)
269. GOST 12020-72, *Plastics. Determination Methods for Resistance to Impact of Chemical Media*, Moscow, Izd. Standartov, 1972. (Rus)
270. *GOST 2789-73. Surface Roughness. Parameters and Characteristics*, Moscow, Izd. Standartov, 1973. (Rus)
271. Ilichev V.D., *Control of Behavior of Birds*, Moscow, Nauka, 1984, 303 p. (Rus)
272. Naumov N.P., *Ecology of Vertebrates*, Moscow, 1970, 375 p. (Rus)
273. Emanuel N.M. and Buchachenko A.L., *Chemical Physics of Molecular Degradation and Stabilization of Polymers*, Moscow, Nauka, 1988, 368 p. (Rus)
274. *The Reference Book on Chemical Technology of Linen Fabric Processing*, Moscow, Legkaya Industria, 1973, 403 p. (Rus)
275. *Protection from Corrosion, Aging and Biodamages of Machines, Equipment and Facilities, The Reference Book*, Moscow, Mashinostroenie, 1987, 784 p. (Rus)
276. Ruban G.I. and Reutova Z.A., Microscopic fungi damaging plastics, *Mikrobiologia i Fitopathologia*, 1976, vol. **10**(5), pp. 238 - 245. (Rus)
277. Anisimov A.A., Feldman M.S., and Vyisotskaya L.B., Micellar fungus enzymes as aggressive metabolites, In Coll.: *Biodamages in Industry*, Gorky, 1985, pp. 8 - 19. (Rus)
278. Munblit V.Ya., Talrose V.L., and Troimov V.I., *Thermal Inactivation of Microorganisms*, Moscow, Nauka, 1985, 248 p. (Rus)
279. Panchenkov G.M. and Lebedev V.P., *Chemical Kinetics and Catalysis*, Moscow, Khimia, 1974, 591 p. (Rus)
280. Emanuel N.M. and Knorre A.G., *The Course of Chemical Kinetics*, Moscow, Vyisshaya Shkola, 1962, 414 p. (Rus)
281. GOST 9.803-88, *The Uniform System of Corrosion and Aging Protection. Fungicides. Effeciency Determination Method*, Moscow, Izd. Standartov, 1989. (Rus)
282. Semenov S.A., Gerasimenko A.A., Chepenko B.A., Ryizhkov A.A., Moiseev Yu.V., Gumargalieva K.Z., Mironova, and Malama A.A., *Patent No. 1,281,591 (USSR) SU A1. A Support for Microbiological Tests*, September 08, 1986. (Rus)
283. Shcherba N.D., Mikitishin S.I., Tyinnyi A.N., and Bartenev G.M., On the cracking mechanism of poly(methyl methacrylate) in media, *Fiz.-Khim. Mekh. Mater.*, 1969, vol. **5**(4), pp. 473 - 478. (Rus)

284. Soshko A.I., On the effect of intermolecular interaction on strength of glassy-like polymers, *Fiz.-Khim. Mekh. Mater.*, 1971, vol. 7(2), pp. 39 – 45. (Rus)
285. Efimov A.V., Bondarev V.V., Kozlov P.V., and Bakeev N.F., The effect of chemical nature of liquid plasticizing media on mechanical properties of crystalline polymers, In: *Proc. Scientific and Technical Conference on Polymer Plasticization*, Kazan, 1980, pp. 53 - 54. (Rus)
286. Mikityuk O.A., Koval V.F., Sashko A.I., Tyinnyi A.N., Perov B.V., and Gudimov M.M., The effect of liquid media of different polarity on mechanical characteristics of polymers, *Fiz.-Khim. Mekh. Mater.*, 1968, vol. **3**(3), pp. 37 - 44. (Rus)
287. Sinevich E.A., Ogorodov R.P., and Bakeev N.F., Adsorptive effect of liquid media on mechanical properties of polymers, *Doklady AN SSSR*, 1973, vol. **212**(6), pp. 1383 - 1385. (Rus)
288. Ryizhkov A.A., Sinevich E.A., Valiotti N.N., and Bakeev N.F., Study of adsorptive effect regularities of liquid media on mechanical properties of polymers, *Vysokomol. Soedin.*, 1976, vol. **616**(3), pp. 212 - 214. (Rus)
289. Voyutski S.S., *The Course of Colloid Chemistry*, Moscow, Khimia, 1976, 511 p. (Rus)
290. Emelianov D.N., Chernorukova Z.G., and Novospasskaya N.Yu., Biocidal composite polyacrylate materials, In: *Mechanics of Composite Materials and Structures*, 1997, vol. **3**(4), pp. 36 - 41. (Rus)
291. Kambour R.P., Structure and properties of crazes in polycarbonate and other glassy polymers, *Polymer*, 1964, vol. **5**(3), pp. 143 – 155.
292. Kambour R.P. and Holik A.S., Electron microscopy of crazes in glassy polymers: use of reinforcing impregnants during microtomy, *J. Polym. Sci.*, 1969, vol. **A-2**(7), pp. 1393 – 1403.
293. Legrand D.G., Kambour R.P., and Haaf W.R., Low-angle X-ray scattering from grazes and fracture surfaces in polystyrene, *J. Polym. Sci.*, 1972, vol. **A-2**(8), pp. 1565 - 1574.
294. Verheulpen-Heymans N., Mechanism of craze thickening during craze growth in polycarbonate, *Polymer*, 1979, vol. **20**(3), pp. 356 - 362.
295. Duckett R.A., Goswani B.C., and Word J.M., Deformation band studies of oriented polyethyleneand poly(ethylene terephthalate), *J. Polym. Sci. - Phys.*, 1977, No. 15, pp. 333—353.
296. *Water in Polymers*, Ed. S.P. Rowland Am. Chem. Soc., Washington, D.C., 1980, 557 p.
297. Anisimov A.A., Semicheva A.S., Vishnyakova T.A., Smirnov V.F., Lyubavina N.P., and Petrova N.P., The effect of some fungicides on respiration intensity and acid formation in Aspergillus niger, *Mikologia i Fitopatologia*, 1980, vol. **14**(1), pp. 24 - 27. (Rus)

298. Smirnov V.F., Feldman N.S., Tolmacheva R.N., and Tarasova N.A., The effect of fungicides on Aspergillus niger respiration intensity and activity of catalase and peoxidase enzymes, In: *Biochemistry and Biophysics of Microorganisms*, Gorky, 1976, pp. 70 - 72. (Rus)
299. *Practical Training on Colloid Chemistry and Electron Microscopy*, Ed. S.S. Voyutski and R.M. Panich, Moscow, Khimia, 1974, 224 p. (Rus)
300. Anisimov A.A., Smirnov V.F., and Semicheva A.S., Biochemical grounds of fungal resistance of polymeric materials, In: *Microorganisms and the Lowest Plants – Degraders of Materials and Articles*, Moscow, 1979, pp. 16 - 27. (Rus)
301. Zaikina N.A. and Duganova N.V., Formation of organic acids separated from biocorroded objects, *Mikologia i Fitopatologia*, 1975, No. 9, pp. 303 - 307. (Rus)
302. Gerasimov V.I., Structural mechanisms of crystalline polymer plastic deformation, *Doctor Dissertation on Chemical Sciences, Thesis*, Moscow, 1980, 238 p. (Rus)
303. Nadareishvili L.I. and Lobzhanidze V.V., On the mechanism of non-polar liquid effect on polyethylene deformability, *Fiz.-Khim. Mekh. Mater.*, 1974, vol. **10**(6), pp. 75 - 79. (Rus)
304. Kazakevich S.A., Kozlov P.V., and Pisarenko A.P., The effect of media and static loads on spontaneous orientation of polymeric films, *Fiz.-Khim. Mekh. Mater.*, 1968, vol. **4**(5), pp. 585 - 590. (Rus)
305. Kazakevich S.A., Kozlov P.V., and Pisarenko A.P., On the reasons of extreme change in strength properties of polymeric films under water effect, *Fiz.-Khim. Mekh. Mater.*, 1969, vol. **5**(1), pp. 75 - 79. (Rus)
306. Kambour R.P., Gruner C.L., and Romagosa E.E., Solvent crazing of "dry" polystyrene and "dry" crazing of plasticizer polystyrene, *J. Polym. Sci. - Phys.*, 1973, vol. **11**(10), pp. 1879 - 1883.
307. Kazakevich S.A. and Kozlov P.V., The effect of preliminary loading on strength of polymeric films, *Fiz.-Khim. Mekh. Mater.*, 1972, vol. **8**(5), pp. 85 - 87. (Rus)
308. Peterlin A., Fracture mechanism of drawn oriented crystalline polymers, *J. Macromol. Sci.*, 1973, vol. **B7**(4), pp. 705 – 727.
309. Peterlin A., Plastic deformation of polymers with fibrous structure, *Colloid and Polym. Sci.*, 1975, vol. **232**(10), pp. 809 – 823.
310. Peterlin A., Morphology and fracture of drawn crystalline polymers, J. Macromol., Sci., 1973, vol. B8(1), pp. 83 - 99.
311. Brown H.R., A theory of the environmental stress cracking of polyethylene, *Polymer*, 1978, No. 19, pp. 1186 – 1188.

312. Soshko A.I., About reciprocal type of liquid medium effect on poly(methyl methacrylate) strength, *Vysokomol. Soedin.*, 1971, vol. **613**(8), pp. 587 – 589.
313. Valyinskaya A.L., Gorokhovskaya T.E., Shitov I.A., and Bakeev N.F., Orientation of low-molecular substances included in polymers, deformed in an adsorption-active medium, *Vysokomol. Soedin.*, 1980, vol. **622**(7), pp. 483 - 484. (Rus)
314. Bandyopadhaya S. and Brown H.R., Environmental stress cracking and morphology of polyethylene, *Polymer*, 1978, No. 19, pp. 589 – 592.
315. Okudu S., *Proc. 11-th Japan Congr. Mat. Res.*, Tokyo, 1967; Kyoto, 1968, pp. 124 – 130.
316. Azhogin F.F., *Corrosive Cracking and Protection of High-Strength Steels*, Moscow, Metallurgia, 1974, 256 p. (Rus)
317. Shluger M.A., Azhogin F.F., and Efimov E.A., *Corrosion and Protection of Metals*, Moscow, Metallurgia, 1981, 215 p. (Rus)
318. *Electrical Properties of Polymers*, Ed. B.I. Sazhin, Leningrad, Khimia, 1986, 224 p. (Rus)
319. Tager A.A. and Tsilipotkina M.V., Porous structure of polymers and sorption mechanism, *Uspekhi Khimii*, 1978, vol. **47**, pp. 152 - 175. (Rus)
320. Stepanov R.D. and Shlenski O.F., *Strength Calculation of Constructions from Plastics Operated in Liquid Media*, Moscow, Mashinostroenie, 1981, 136 p. (Rus)
321. Minsker K.S. and Fedoseeva G.T., *Degradation and Stabilization of Polyvinylchloride*, 2nd Ed., Moscow, Khimia, 1979, 271 p. (Rus)
322. Minsker K.S., Kolesov S.V., and Zaikov G.E., *Aging and Stabilization of Vinyl Chloride-Based Polymers*, Moscow, Nauka, 1982, 272 p. (Rus)
323. Rudakova A.K., Microbial corrosion of polymeric materials used in cable industry, In: *Problems of Biological Damages and Overgrowths*, Moscow, 1972, pp. 32 - 44. (Rus)
324. Tarasova N.A., Smirnov V.F., and Drinko Z.N., Study of biodamages of materials used in radio engineering, In: *Biodamages in Industry*, Gorky, 1983, pp. 52 - 57. (Rus)
325. Minsker K.S., Berlin Al.Al., Lisitski V.V., and Kolesov S.V., Mechanism and kinetics of polyvinylchloride dehydrochlorination, *Vysokomol. Soedin.*, 1977, vol. **A19**(1), pp. 32 - 36. (Rus)
326. Smith A., *Analytical Chemistry of Polymers*, P. II, Interscience, 1976.
327. Minsker K.S., Abdullin M.I., and Kalashnikov V.G., Oxidative thermodegradation of plasticized PVC, *Vysokomol. Soedin.*, 1980, vol. **22**(9), pp. 2131 - 2132. (Rus)

328. Lyitkina N.I., Egorova T.M., and Mizerovski A.N., Direct detection of plasticizers in PVC by UV-spectrophotometry method, *Plastmassy*, 1983, No. 3, p. 58. (Rus)
329. Rabek Y., *Newer Methods of Polymer Characterization*, Interscience, 1964.
330. Bezveliev B.M., Plasticizer migration from poly(vinyl chloride) plasticate, Plastmassy, 1967, No. 2, pp. 8 - 9. (Rus)
331. Borisov B.I., Low-molecular component extraction from insulating coverings, *Colloid. Zh.*, 1973, vol. **35**(1), p. 140. (Rus)
332. Borisov B.I., Estimation of the factor of plasticizer migration from PVC in soils, *Colloid. Zh.*, 1978, vol. **40**(3), pp. 535 - 553. (Rus)
333. Borisov B.I. and Gromov N.I., The effect of plasticizer migration on properties of underground pipeline coatings, In: *Corrosion and Protection in Oil and Gas Industry*, RNTS VNIIOEHG, 1977, No. 3, pp. 19 - 22. (Rus)
334. Kazakyavichute G.A., Korchagin M.V., and Kutieva O.A., Imparting anti-microbial properties to cellulose textile materials using nitrofuran sequence derivatives, *Izvestiya VUZov. Tekhnologia Tekstilnoi Promyishlennosti*, 1985, No. 6, pp. 59 - 62. (Rus)
335. Kolontarov I.Ya. and Liverant V.L., *Imparting Biocidal Properties and Resistance to Microorganisms to Textile Materials*, Dushanbe, Donish, 1981, 202 p. (Rus)
336. Karpukhina S.Ya., Gumargalieva K.Z., Daurova T.T., and Mironova V.A., Fermentative degradation of polyglycolide, *Vysokomol. Soedin.*, 1983, vol. **B15**(12), pp. 209 - 211. (Rus)
337. Voronkova O.S., Daurova T.T., Moiseev Yu.V., and Gumargalieva K.Z., PET degradation under conditions of concomitant infection, *Vysokomol. Soedin.*, 1981, No. 12, pp. 325 – 331. (Rus)
338. Tomashov N.D. and Chernova G.P., *Corrosion Theory and Corrosion Resistant Construction Alloys*, Moscow, Metallurgia, 1986, 359 p. (Rus)
339. Sinyavski V.S., Valkov V.D., and Budov G.N., *Corrosion and Protection of Aluminum Alloys*, Moscow, Metallurgia, 1979, 224 p. (Rus)
340. Mikhailovski Yu.N., Theory and practice of atmospheric corrosion rate calculation from meteorological characteristics, *Scientific-Technical Conference on the Problem of "Development of Measures Protecting Metals from Corrosion"*, Moscow, 1971, pp. 111 - 114. (Rus)
341. Berukshtis G.K. and Clark G.B., *Corrosive Resistance of Metals and Metal Coverings under Atmospheric Conditions*, Moscow, Nauka, 1971, 275 p. (Rus)

342. Panchenko Yu.M., Shuvakhina L.A., and Mikhailovski Yu.M., Atmospheric corrosion of metals in the Far East region, *Protection of Metals*, 1982, vol. **XVIII**(4), pp. 354 - 356. (Rus)
343. Panchenko Yu.M. and Shuvakhina L.A., Dependence of atmospheric corrosion rate of metals on climatic conditions of the Far East region, *Protection of Metals*, 1984, vol. **XX**(6), pp. 16 - 18. (Rus)
344. Semenov S.A., Skribachilin V.B., and Zhdanova O.A., Testing of machinery cleaning means against microbiological pollutions. Methodological guidelines, *Vyipusk VVS No. 6440*, Vedomstvennoe Izdatelstvo, 1990, 36 p. (Rus)
345. Kacharov S.A., Semenov S.A., Sukova O.I., and Il'in A.A., Restoration of resistance of articles from cellulose-containing fabrics to biodamages under operation and storage conditions of machinery, *Vyipusk VVS No. 6389*, Vedomstvennoe Izdatelstvo, 1991, 32 p. (Rus)
346. Semenov S.A., Gerasimenko A.A., and Gumargalieva K.Z., Estimation of biocide efficiency for polymeric materials used in machinery. Methodological guidelines, *Vyipusk VVS No. 5392*, Vedomstvennoe Izdatelstvo, 1985. (Rus)
347. Matyusha G.V., Semenov S.A., and Mikhailova L.K., Catalogue of microorganisms-degraders of weapons and armament. Part 1. Collection of the Ministry of Defense of the Russian Federation, *Vyipusk VVS No. 6973*, Vedomstvennoe Izdatelstvo, 1996, 56 p. (Rus)
348. Matyusha G.V., Mikhailova L.K., and Semenov S.A., Catalogue of microorganisms-degraders of weapons and armament. Part 2. Collection of the Ministry of Defense of the Russian Federation, *Vyipusk VVS No. 7018*, Vedomstvennoe Izdatelstvo, 1997, 27 p. (Rus)
349. Matyusha G.V., Semenov S.A., and Mikhailova L.K., Catalogue of microorganisms-degraders of weapons and armament. Part 3. Collection of the Ministry of Defense of the Russian Federation, *Vyipusk VVS No. 7069*, Vedomstvennoe Izdatelstvo, 1999, 43 p. (Rus)
350. Gerasimenko A.A., Semenov S.A., Chepenko B.A., Moiseev Yu.V., Gumargalieva K.Z., Mironova S.N., Malama A.A., and Ryizhkov A.A., A method of polymeric material bioresistance estimation, *Patent SU 1,331,268 A (USSR)*, April 15, 1987. (Rus) November 01, 1988. (Rus)
351. Semenov S.A., Ryizhkov A.A., Chepenko B.A., Gerasimenko A.A., Moiseev Yu.V., and Gumargalieva K.Z., A method of polymer fungal resistance determination, *Patent SU 1,462,998 (USSR)*, November 01, 1988. (Rus)
352. Semenov S.A., Fidler Kh.N., Zimbakhadze D.V., Khachaturova O.A., Moiseev Yu.V., and Gumargalieva, A method of applying water-insoluble

biocide to surface of a hydrogel support, *Patent SU 1,577,364 (USSR)*, March 08, 1990. (Rus)

353. Matyusha G.V., Semenov S.A., Kartasheva T.A., Arnautova V.A., and Sizova T.P., Penicillium rubrumstoll fungus culture used as a test-culture for determination of microbiological resistance of conservation lubricants, applied in tropical climate, *Patent RU 2,045,139 C2 (Russian Federation)*, September 27, 1995. (Rus)
354. Matyusha G.V., Semenov S.A., Kartasheva T.A., and Arnautova V.A., Bacillus sp. bacterium culture used as a test-culture for determination of bacteriological resistance of steel alloys, *Patent RU 2,080,375 C1 (Russian Federation)*, May 27, 1997. (Rus)
355. Matyusha G.V., Semenov S.A., Svitich A.A., Kartasheva T.A., Blinkova L.P., and Butova L.G., Bacillus sp. bacterium culture used as a test-culture for determination of bacteriological resistance of aluminum and magnesium alloys, *Patent RU 2,080,374 C1 (Russian Federation)*, May 27, 1997. (Rus)
356. Matyusha G.V., Semenov S.A., Starikov N.E., Pavlov V.N., and Blinkova L.P., Bacillus sp. bacterium culture – the producer of a biocide of tropical micromicetes-degraders of technical oils, *Patent RU 2,141,523 C1 (Russian Federation)*, November 20, 1999. (Rus)
357. Chuev Yu.V. and Mikhailov Yu.B., *Forecasting in Armament*, Moscow, Voenizdat, 1975, 279 p. (Rus)
358. *Reference Books on Climate of the USSR*, vols. **1 - 34**, Parts IV, V, Leningrad, Gidrometeoizdat, 1968. (Rus)
359. Golubev A.N. and Kadyirov M.Kh., *Forecasting of Metal Corrosion under Atmospheric Conditions*, Moscow, GOSINTI, 1967, No. 3, 67 – 487/13. (Rus)
360. Kachurin L.G., *Electrical Measurements of Air-Physical Values*, Moscow, Vyisshaya Shkola, 1967. (Rus)
361. Baigozhin A.A., Kuznetsova L.N., Semenov S.A., Nikolaev E.M., and Sukova O.I., A method of protection of optical component surfaces from biodamages, *Patent RU 2,111,182 C1 (Russian Federation)*, May 20, 1998. (Rus)
362. Borisova I.A., Kryuchkov A.A., Semenov S.A., Vorobiev B.P., Zhdanova O.A., and Fedin A.N., Fungus-resistant varnish composite, *Patent RU 2,111,996 C1 (Russian Federation)*, May 27, 1998. (Rus)
363. Il'in A.A., Kacharov S.A., and Semenov S.A., A preparation for restoration of operation properties of articles from cellulose-containing fabrics for special purposes, *Patent RU 2,091,528 C1 (Russian Federation)*, September 27, 1997. (Rus)

364. Nazarov V.G., Stolyarov V.P., Dedov A.V., Semenov S.A., and Manin V.N., A method of surface modification of crystalline and amorphous thermoplasts and rubbers, *Patent SU 1,816,773 A1 (USSR)*. (Rus)
365. Semenov S.A., Zhdanova O.A., Babkin V.G., and Sukova O.I., A method of modification of polymeric article surfaces, *Patent SU 1,684,084 A1 (USSR)*, June 15, 1991. (Rus)
366. Kuzinova T.N., Altsibeeva A.I., Semenov S.A., and Kalinovski S.A., An inhibitor of atmospheric and biological corrosion of metals, *Patent RU 2,083,719 C1 (Russian Federation)*, June 10, 1997. (Rus)
367. Matyusha G.V., Semenov S.A., and Blinkova L.P., A protection method of technical lubricants from bacterium impact, *Patent RU 2,143,200 C1 (Russian Federation)*, December 27, 1999. (Rus)
368. Gerasimenko A.A., Semenov S.A., Chepenko B.A., Kacharov S.A., and Il'in A.A., Fungistatic, *Patent SU 1,367,181 A (USSR)*, September 15, 1987. (Rus)
369. Semenov S.A., Gerasimenko A.A., Chepenko B.A., Ryizhkov A.A., Gumargalieva K.Z., Moiseev Yu.V., Malama A.A., and Mironova S.N., Fungistatic for protection of metal surfaces and coverings, *Patent SU 1,350,855 A1 (USSR)*, July 08, 1987. (Rus)
370. Chernin Yu.V., Erokhina A.A., Semenov S.A., Alekseeva T.N., and Sukova O.I., Cleansing and disinfecting compound for treating metal surface, *Patent RU 2,084,498 C1 (Russian Federation)*, July 20, 1997. (Rus)
371. Semenov S.A., Zhdanova O.A., Sukova O.I., Osminov V.A., and Soloviev A.I., Cleansing and disinfecting compound for purifying a surface from pollutions, *Patent SU 1,807,077 A1 (USSR)*, October 10, 1992. (Rus)
372. Andreyuk E.I., Bilai V.I., Koval E.Z., and Kozlova I.A., *Microbial Corrosion and Its Exciters*, Kiev, Naukova Dumka, 1980, 287 p. (Rus)
373. Ashmarin I.P. and Vorobiev A.A., *Statistic Methods in Microbiological Investigations*, Moscow, Medgiz, 1982, 180 p. (Rus)
374. Ivanushkina N.E. and Mirchink T.G., The use of radial growth rate of micromycetes as ecological index, *Mikrobiologia*, 1982, vol. **52**(6), pp. 941 – 944. (Rus)
375. Kanevskaya G.A., *Biological Damaging of Industrial Materials*, Leningrad, Nauka, 1984, 230 p. (Rus)
376. Kochkina G.A., Mirchink T.G., Kozhevin P.A., and Zvyagintsev D.G., The radial growth rate of fungi colonies connected with their ecology, *Mikrobiologia*, 1978, vol. **47**(5), pp. 964 – 965. (Rus)
377. Malama A.A., Mironova S.N., Filimonova T.V. et al., The effect of temperature on growth of mycelial fungi colonies, *Mikrobiologia*, 1985, vol. **54**(6), pp. 994 - 997. (Rus)

378. *Method of Experimental Mycology, The Reference Book*, Ed. V.I. Bilai, Kiev, Naukova Dumka, 1982, 550 p. (Rus)
379. Panikov N.S., Afremov V.D., and Aseeva I.V., Kinetics of growth of *Mucor pluembeus* and *Mortierella ramanniana* colonies on agarized media with glucose, *Mikrobiologia*, 1981, vol. **50**(1), pp. 55 – 61. (Rus)
380. Pert S.D., *Grounds for Microorganisms and Cells Cultivation*, Moscow, Mir, 1978, 331 p. (Rus)
381. Rabotnova I.L., Some data on regularities of microorganism growth, *Zh. Obshch. Biologii*, 1972, vol. **33**(5), pp. 533 - 554. (Rus)
382. Smirnov V.F. and Feldman M.S., Study of the radial growth rate of *Aspergillus flavus* colonies on various polymeric substrates, In: *Microorganisms – Producers of Biologically Active Substances*, Riga, AN LatvSSR, 1984, p. 95. (Rus)
383. Tsyipkin A.G., *The Reference Book on Mathematics*, Moscow, Nauka, 1983, 480 p. (Rus)
384. Shlegel G., *General Mikrobiologia*, Moscow, Mir, 1972, pp. 178 - 185. (Rus)
385. Trilli A., Constanzi I., Lamanna F., and Di Dio N., Development of agar-disc method for the rapid screening of strains with increased productivity, *J. Chem. Technol. Biotechnol.*, 1982, vol. **32**(1), pp. 281 – 291.
386. Trinci A., A kinetic study of the growth of *Aspergillus nidulans* and other fungi, *J. Gen. Microbiol.*, 1969, vol. **57**, pt. 1, pp. 11 – 24.
387. Trinci A., Influence of the width of the peripheral growth zone on the radial growth rate of fungal colonies on solid media, *Ibid.*, 1971, vol. **67**, pt. 3, pp. 325 – 344.
388. Anisimon A.A., Smirnov V.F., and Semicheva A.S., Biochemical grounds of fungal resistance of polymeric materials, In: *Microorganisms and the Lowest Plants – the Destructors of Materials and Articles*, Moscow, Nauka, 1979, pp. 16 – 21. (Rus)
389. Zvyagintsev D.G., *Soil Microbiology Methods*, Moscow, Izd. MGU, 1980, pp. 78 – 79. (Rus)
390. Ivanushkina N.E. and Mirchink T.G., The use of radial rate of micromycetes growth as the ecological index, *Mikrobiologia*, 1982, vol. **51**(6), pp. 941 - 944. (Rus)
391. Kochkina G.A., Mirchink T.G., Kozhevin P.A., and Zvyagintsev D.G., Radial rate of fungus colonies growth associated with their ecology, *Mikrobiologia*, 1978, vol. **47**(5), pp. 964 - 965. (Rus)
392. *Physik*, H. Kuchling, VEB Fachbuch Verlag, 1980.
393. Malama A.A., Mironova S.N., Filimonova T.V., Moiseev Yu.V., and Gumargalieva K.Z., The effect of temperature on mycelial micromycetes colonies growth, *Mikrobiologia*, 1985, vol. **54**(6), pp. 994 - 997. (Rus)

394. Malama A.A., Shinkarchuk V.N., and Lukashik A.N., The effect of pentachlorophenol and sulfanol on radial growth of *Penicillium chrysogenum* Thom colonies, Mikologia i fitopatologia, 1987a, vol. **21**(4), pp. 330 - 336. (Rus)

395. Malama A.A., Mironova S.N., Filimonova T.V., Semenov S.A., and Ryizhkov A.A., Kinetics of the radial growth of Aspergillus colonies at different temperatures, *Mikrobiol. Zh.*, 1987b, vol. **49**(2), pp. 46 – 49. (Rus)

396. Panikov N.S., Afremov V.D., and Aseeva N.V., Kinetics of *Mucor plumbeus* and *Mortierella ramanniana* colonies growth on agarized media with glucose, *Mikrobiologia*, 1982, vol. **51**(6), pp. 941 - 944. (Rus)

397. *The Reference Book on Microbiological and Virological Investigation Methods*, Ed. M.O. Birger, Moscow, Meditsina, 1982, p. 44. (Rus)

398. Turkova Z.A. and Fomina N.V., Properties of *Aspergillus penicilloides* damaging optical articles, *Mikologia i Fitopatologia*, 1982, vol. **16**(4), pp. 314 - 317. (Rus)

399. Chebotarev L.N. and Generalova N.A., Kinetics of physiological and biochemical processes in cultivation of *Aspergillus niger* van Thiegh. on lactic serum, *Mikologia i Fitopatologia*, 1984, vol. **18**(2), pp. 135 - 138. (Rus)

400. Gooday G.W., Anautoradiographic study of hyphal growth of some fungi, *J. Gen. Microbiol.*, 1971, vol. **67**(2), pp. 125 – 133.

401. Robson G.D., Bell S.D., Kuhn P.J., and Trinci A.P., Glucose and penicillinum concentration in agar medium below fungal colonies, *J. Gen. Microbiol.*, 1987, vol. **133**(2), pp. 361 – 367.

402. Trinci A.P., A kinetic study of the growth of *Aspergillus niger* and other fungi, *J. Gen. Microbiol.*, 1969, vol. **57**(1), pp. 11 – 34.

403. Trinci A.P., Influence of the width of the peripheral growth zone on the radial growth rate of fungal colonies on solid media, *J. Gen. Microbiol.*, 1971, vol. **67**(3), pp. 325 – 344.

404. Trilli A., Constanzi I., Lamanna F., and Di Dio N., Development of the agar-disc method for the rapid screening of strains with increased productivity, *J. Chem. Technol. Biotechnol.*, 1982, vol. **32**(1), pp. 281 – 291.